MW00752380

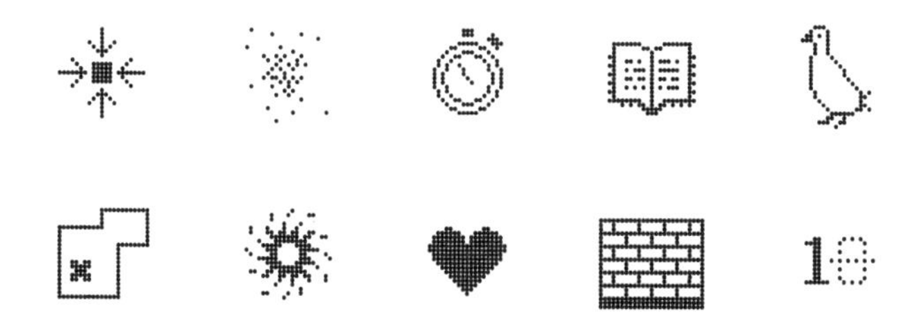

THE LAWS OF SIMPLICITY

John Maeda

DESIGN, TECHNOLOGY, BUSINESS, LIFE

The MIT Press
Cambridge, Massachusetts
London, England

MIT Press books may be purchased at special quantity discounts for business or sales promotional use. For information, please email special_sales@mitpress.mit.edu or write to Special Sales Department, The MIT Press, 55 Hayward Street, Cambridge, MA 02142.

This book was set in Mercury Text and DotMatrix-Two by the author and was printed and bound in the United States of America.

Library of Congress Cataloging-in-Publication Data

Maeda, John.
The laws of simplicity / John Maeda.
p cm.—(Simplicity: Design, Technology, Business, Life)
Includes index.
ISBN 0-262-13472-1—ISBN 978-0-262-13472-9 (hc. : alk. paper)
1. Systems engineering. I. Title.
TA168 .M255 2006 650.1—dc22 2006044885

10 9 8 7 6 5 4 3

For Kris

I promise to love you more, and never less.

Contents

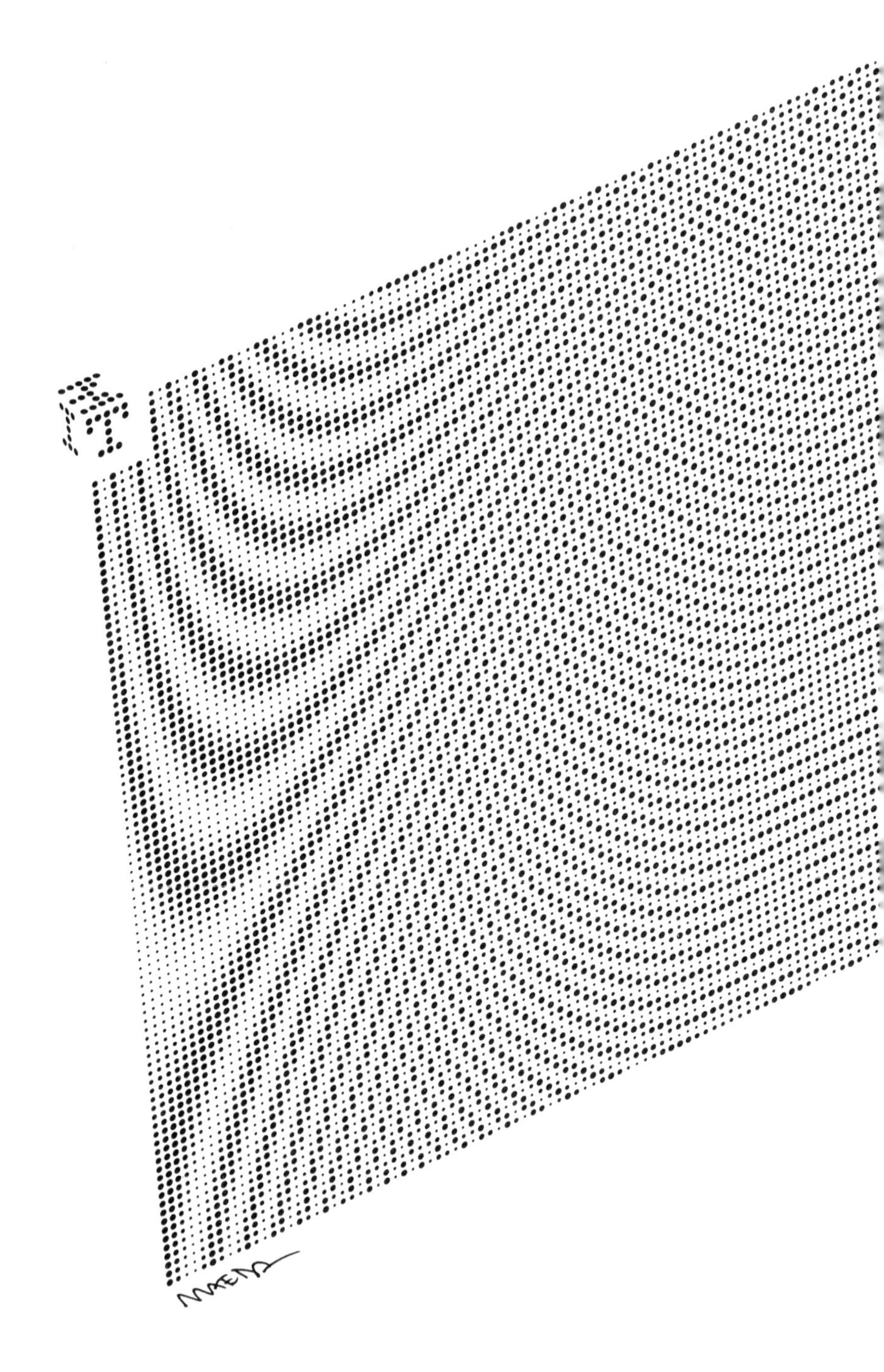

SIMPLICITY = SANITY

Technology has made our lives more full, yet at the same time we've become uncomfortably "full."

I watched the process whereby my daughters gleefully got their first email accounts. It began as a tiny drop—emails sent among themselves. It grew to a slow drip as their friends joined the flow of communication. Today it is a waterfall of messages, e-cards, and hyperlinks that showers upon them daily.

I urge them to resist the temptation to check their email throughout the day. As adults, I tell them, they will have ample opportunity to swim in the ocean of information. "Stay away!" I warn, because even as an Olympic-class technologist, I find myself barely keeping afloat. I know that I'm not alone in this feeling of constantly drowning—many of us regularly engage (or don't) in hundreds of email conversations a day. But I feel somewhat responsible.

My early computer art experiments led to the dynamic graphics common on websites today. You know what I'm talking about—all that stuff flying around on the computer screen while you're trying to concentrate—that's me. I am partially to blame for the unrelenting stream of "eye candy" littering the information landscape. I am sorry, and for a long while I have wished to do something about it.

Achieving simplicity in the digital age became a personal mission, and a focus of my research at MIT. There, I straddle the fields of design, technology, and business as both educator and practitioner. Early in my ruminations I had the simple observation that the letters "M," "I," and "T"—the letters by which my university is known—occur in natural sequence in the word SIMPLICITY. In fact, the same can be said of the word COMPLEXITY. Given that the "T" in M-I-T stands for "technology"—which is the very source of much of our feeling overwhelmed today—I felt doubly responsible that someone at MIT should take a lead in correcting the situation.

In 2004, I started the MIT SIMPLICITY Consortium at the Media Lab, comprised of roughly ten corporate partners that include AARP, Lego, Toshiba, and Time. Our mission is to define the business value of simplicity in communication, healthcare, and play. Together we design and create prototype systems and technologies that point to directions where simplicity-driven products can lead to market success. By the publication date of this book, a novel networked digital photo playback product co-developed with Samsung will serve as an important commercial data point to test the validity of the Consortium's stance on simplicity.

When the blogosphere began to emerge, I responded and created a blog about my evolving thoughts on simplicity. I set out to find a set of "laws" of simplicity and targeted sixteen principles as my goal. Like most blogs, it has been a place where I have shared unedited thoughts that represent my personal opinions on a topic about which I am passionate. And although the theme of the blog began just along the lines of design, technology, and business I discovered that the readership resonated with the topic that underlies it all: my struggle to understand the meaning of life as a humanist technologist.

Through my ongoing journey I've discovered how complex a topic simplicity really is, and I don't pretend to have solved the puzzle. Having recently spoken to an 85-year old MIT linguistics professor who has been working on the same problem his entire life, I am inspired to grapple with this puzzle for many more years. My blog led me to the fact that there aren't sixteen laws, but rather the ten published in this volume. Like all man-made "laws" they do not exist in the absolute sense—to break them is no sin. However you may find them useful in your own search for simplicity (and sanity) in design, technology, business, and life.

SIMPLICITY AND THE MARKETPLACE

The marketplace abounds with promises of simplicity. Citibank has a "simplicity" credit card, Ford has "keep it simple pricing," and Lexmark vows to "uncomplicate" the consumer experience. Widespread calls for simplicity formed a trend that was inevitable, given the structure of the technology business around selling the same thing "new and improved" where often "improved" simply means *more*. Imagine a world in which software companies simplified their programs every year by shipping with 10% fewer features at 10% higher cost due to the expense of simplification. For the consumer to get less and pay more seems to contradict sound economic principles. Offer to share a cookie with a child and which half will the child want?

Yet in spite of the logic of demand, "simplicity sells" as espoused by *New York Times* columnist David Pogue in a presentation at the 2006 annual TED Conference in Monterey. The undeniable commercial success of the Apple iPod—a device that does less but costs more than other digital music players—is a key supporting example of this trend. Another example is the deceivingly spare interface of the powerful Google search engine, which is so popular that "googling" has become shorthand for "searching the Web." People not only buy, but more importantly love, designs that can make their lives simpler. For the foreseeable future, complicated technologies will continue to invade our homes and workplaces, thus simplicity is bound to be a growth industry.

Simplicity is a quality that not only evokes passionate loyalty for a product design, but also has become a key strategic tool for businesses to confront their own intrinsic complexities. Dutch conglomerate Philips leads in this area with its utter devotion to realizing "sense and simplicity." In 2002 I was invited by Board of Management Member Andrea Ragnetti to join Philips' "Simplicity Advisory Board (SAB)." I initially thought that "sense and simplicity" was merely a branding effort, but when I met in Amsterdam with Ragnetti and his CEO Gerard Kleisterlee at the first meeting of the SAB I saw the greater ambition. Philips plan to reorganize not only all of their product lines, but also their entire set of business practices around simplicity. When I tell this story to industry leaders the consistent feedback I get is that Philips is not alone in the quest to reduce the complexities of doing business. The hunt is on for simpler, more efficient ways to move the economy forward.

WHOM IS THIS BOOK FOR?

As an artist, I'd like to say that I wrote this book for myself in the spirit of climbing a mountain "because it's there." But the reality is that I wrote it in response to the many voices of encouragement—either by email or in person—from people that wish to better understand *simplicity*. I've heard from biochemists, production engineers, digital artists, homemakers, technology entrepreneurs, road construction administrators, fiction writers, realtors, and office workers, and the interest just seems to keep on growing. With support there is always discouragement: some worry about the negative connotations of simplicity where it can lead to a simplistic and "dumbed-down" world. You will see in the latter part of this book that I position complexity and simplicity as having importance relative to each other as necessary rivals. Thus I realize that although the idea of ridding the earth of complexity might seem the shortest path to universal simplicity, it may not be what we truly desire.

I originally conceived this book as a sort of Simplicity 101, to give readers an understanding of the foundation of simplicity as it relates to design, technology, business, and life. But now I see that a foundation can wait until I'm 85 like my professor friend, and for now a framework will suffice which you now hold in your hands. Also, in the course of completing my MBA, I found that the majority of books on innovation and business are published by a single authority. I have been mellowed by many sobering events in my otherwise extremely fortunate life, so I was looking for something that was more heartful than a book specifically aimed at the technology or business market.

My good friends at the MIT Press were supportive of a softer and more creative approach to the developing arena of simplicity and here you have the first step in such a series. The price-point and design of these books were carefully targeted for the distinguishing reader that is looking for something new and different. At the heart of the series is a focus on the business of technology, grounded in an expert's knowledge of design, and with a light touch of curiosity about life. I welcome you to this creative experience.

HOW-TO USE THIS BOOK

The ten Laws outlined in the body of this book are generally independent of each other and can be used together or alone. There are three flavors of simplicity discussed here, where the successive set of three Laws (1 to 3, 4 to 6, and 7 to 9) correspond to increasingly complicated conditions of simplicity: basic, intermediate, and deep. Of the three clusters, basic simplicity (1 to 3) is immediately applicable to thinking about the design of a product or the layout of your living room. On the other hand, intermediate simplicity (4 to 6) is more subtle in meaning, and deep simplicity (7 to 9) ventures into thoughts that are still ripening on the vine. If you wish to save time (in accordance with the third Law of TIME), I suggest you start with basic simplicity (1 to 3) and then skip to the tenth Law of THE ONE which sums up the entire set.

Each section is a collection of micro-essays that cluster around the main topic presented. Rarely do I have answers, but instead I have a lot of questions just like you. Every Law begins with an icon of my design that represents the basic concepts I present. The images are not a literal explanation of the contents, but may help you to better appreciate each of the Laws. There is also associated Web content at *lawsofsimplicity.com* where you can download the artwork as desktop patterns in case that will help to motivate you.

In addition to the ten Laws, I offer three Keys to achieving simplicity in the technology domain. Think of them as areas in which to invest R&D resources, or simply to keep an eye on. How these Keys, and the Laws, connect to market valuation is a new hobby of mine. Those experiments and further predictions of simplifying technology trends are visible as a free service on *lawsofsimplicity.com* as well.

I intentionally capped the total page count at 100 pages in accordance with the TIME-saving third Law—which is truly dear to my heart. Thus the entire book can be read during your lunch break or else on a short flight. But please don't feel pressured to rush through this book. When I first set out with youthful zeal to attack the simplicity question, I felt that complexity was destroying our world and had to be stopped! At a conference where I later spoke, a 73-year old artist took me aside and said, "The world's *always* been falling apart. So relax." He's probably right. So take his advice and try to LEAN BACK while you read this book, if you can.

ACKNOWLEDGMENTS

I would like to thank Ellen Faran and Robert Prior of the MIT Press for shepherding the process of publishing this book at a speed unlike any other. The appropriateness of simplicity as a concept coming from MIT made immediate sense to both of them from the beginning. Given the support I've experienced from the MIT Press, I know that their enthusiasm was infectious in a way that made a normally complex task get executed more simply. Of course I would not wish it any other way ;-).

The inspirations for this book are many, and most of them are evident throughout the discussion of the Laws. I don't take inspiration lightly—it sits squarely in the middle of my BRAIN, as presented in the fourth Law of LEARN. I continue to look to inspiration from my brilliant graduate students, energetic undergraduates, incredible staff, and unparalleled colleagues at MIT, especially at the Media Lab.

My texts were tuned and simplified by the masterful literary mind of Jessie Scanlon. I've known Jessie since her *Wired Magazine* days and always look to her for the latest information on breaking trends in design. Jessie was my writing Master in this process, and I appreciate her time and patience.

A final pass of meticulous edits was executed by my students Burak Arikan, Annie Ding, Brent Fitzgerald, Amber Frid-Jimenez, Kelly Norton, and Danny Shen. Thank you guys!

Finally, I thank my wife Kris and our daughters for keeping my life both wonderfully complex, yet infinitely simple.

TEN LAWS

1 **REDUCE** The simplest way to achieve simplicity is through thoughtful reduction.

2 **ORGANIZE** Organization makes a system of many appear fewer.

3 **TIME** Savings in time feel like simplicity.

4 **LEARN** Knowledge makes everything simpler.

5 **DIFFERENCES** Simplicity and complexity need each other.

6 **CONTEXT** What lies in the periphery of simplicity is definitely not peripheral.

7 **EMOTION** More emotions are better than less.

8 **TRUST** In simplicity we trust.

9 **FAILURE** Some things can never be made simple.

10 **THE ONE** Simplicity is about subtracting the obvious, and adding the meaningful.

THREE KEYS

1 **AWAY** More appears like less by simply moving it far, far away.

2 **OPEN** Openness simplifies complexity.

3 **POWER** Use less, gain more.

LAW 1

Law 1

REDUCE

The simplest way to achieve simplicity is through thoughtful reduction.

The easiest way to simplify a system is to remove functionality. Today's DVD, for instance, has too many buttons if all you want to do is play a movie. A solution could be to remove the buttons for Rewind, Forward, Eject, and so forth until only one button remains: Play.

But what if you want to replay a favorite scene? Or pause the movie while you take that all-important bathroom break? The fundamental question is, where's the balance between simplicity and complexity?

HOW SIMPLE CAN YOU MAKE IT?	←···→	**HOW COMPLEX DOES IT HAVE TO BE?**

On the one hand, you want a product or service to be easy to use; on the other hand you want it to do everything that a person might want it to do.

The process of reaching an ideal state of simplicity can be truly complex, so allow me to simplify it for you. *The simplest way to achieve simplicity is through thoughtful reduction.* When in doubt, just remove. But be careful of what you remove.

SHE'S ALWAYS RIGHT

We would find it hard to remove any given button from a DVD player if forced to do so. The problem is one of choosing what deserves to live, at the sacrifice of what deserves to die. Such decisions are not easy when most of us are not trained to be despots. Our usual preference is to let live what lives: we would choose to keep all the functionality if we could.

When it is possible to reduce a system's functionality without significant penalty, true simplification is realized. When everything that can be removed is gone, a second battery of methods can be employed. I call these methods **SHE: SHRINK, HIDE, EMBODY.**

SHE: SHRINK

When a small, unassuming object exceeds our expectations, we are not only surprised but pleased. Our usual reaction is something like, "That little thing did all that?" Simplicity is about the unexpected pleasure derived from what is likely to be insignificant and would otherwise go unnoticed. The smaller the object, the more forgiving we can be when it misbehaves.

Making things smaller doesn't make them necessarily better, but when made so we tend to have a more forgiving attitude towards their existence. A larger-than-human-scale object demands its rightful respect, whereas a tiny object can be something that deserves our pity. When comparing a kitchen spoon to a construction bulldozer the larger scale of the vehicle instills fear, while the rounded utensil appears harmless and inconsequential. The bulldozer can run you over and end your life, but if the spoon were to fall on top of you, your life would likely be spared. Guns, mace cannisters, and little karate experts of course are the exception to this rule of "fear the large, endear the small."

Technology is SHRINK-ing. The computational power of a machine that sixty years ago weighed 60,000 pounds and occupied 1,800 square feet can now be packed onto a sliver of metal less than a tenth the size of the nail on your pinkie. Integrated circuit (IC) chip technology—commonly referred to as "computer chips"—allows far greater complexity at a much tinier scale. IC chips lie at the heart of the problem of complex devices today as they enable increasingly smaller devices to be created. A kitchen spoon and a mobile phone can share the exact same physical dimensions, yet the many IC's embedded inside the phone make the device easily more complex than the bulldozer—so looks can be deceiving.

Thus while IC's are a primary driver of complexity in modern day objects, they also enable the ability to shrink a frighteningly complex machine to the size of a cute little gumdrop. The smaller the object is, the lower the expectations; the more IC's that are inside, the greater the power. In this age of wireless technology that connects the IC inside the phone with all the computers in the world, power has now become absolute. There is no turning back to the age when large objects were complex and small objects were simple.

Babies are examples of complex machines that although small, require constant attention to the point of driving most parents insane. Yet in the midst of the havoc they wreak, a precious moment can give way when their big beautiful eyes peer into your tired bleariness with a look of, "Help me! Love me!" It is said that this irresistible cuteness is their key self-preservation mechanism, which I know myself works for a fact, having experienced it many times over. Fragility is an essential counteracting force to complexity because it can instill pity—which by coincidence also occurs in the word SIMPLICITY!

The science of making an object appear delicate and fragile is a skill practiced throughout the history of art. An artist is trained to evoke emotion in his fellow human being through the work he creates, whether that emotion be pity, fear, anger, or any other feeling or combination thereof. Of the many tools at the artist's disposal to achieve enhanced small-ification are lightness and thinness.

For example, the mirrored back of an Apple iPod creates the illusion that the object is only as thin as the floating white or black plastic layer because the rest of the object adapts to its surroundings. Already thin, flat-screen displays like LCD's or plasmas are made to appear even lighter by sitting atop minimal structural supports or in the extreme case floating on an invisible Lucite platform. Another common approach to achieving thinness is seen in the Lenovo ThinkPad's beveled clamshell as your eyes travel down and off the bottom edge of the keyboard to nothingness. A further collection of these types of designs can be browsed at *lawsofsimplicity.com* at your convenience.

Any design that incorporates lightness and thinness conveys the impression of being smaller, lesser, and humbler. Pity gives way to respect when much more value is delivered than originally expected. There is a steady stream of core technologies that will make things smaller, such as nanotechnology—the science of building machines that fit between your squeezed thumb and forefinger. Lessening the inevitable complicating blow of these technologies by way of SHRINK may seem like a form of deception, which it is. But anything that can make the medicine of complexity go down easier is a form of simplicity, even when it is an act of deceit.

SHE: HIDE

When all features that can be removed have been, and a product has been made slim, light, and thin, it's time for the second method: HIDE the complexity through brute-force methods. A classical example of this technique is the Swiss army knife. Only the tool you wish to use is exposed, while the other blades and drivers are hidden.

With an endless array of buttons, remote controls for audio/video equipment are notoriously confusing. In the 90s, a common design solution was to hide the less-used functions, such as setting the time or date behind a hidden door, while keeping only the primary functions such as Play, Stop, and Eject exposed. This approach is no longer popular, probably due to a combination of the added production costs and the prevailing belief that visible features (i.e. buttons) attract buyers.

As style and fashion have become powerful forces in the cell phone market, handset makers have been pushed to find the balance between the elegance of simplicity and need-it-all complexity. Today, the clamshell design is the most evolved example of hiding functionality until you really need it. All buttons are sandwiched between the speaker and microphone such that when it is closed it is a simple bar of soap. Many recent designs have gone beyond the clamshell, and employ slide-away or flip-out mechanisms. Such evolutions are driven by a market that demands innovation and is willing to pay for clever ways to HIDE complexity.

But there might be no better example of the HIDE method than today's computer interfaces. The menu bar at the top hides the functionality of the application. And the other three sides of the screen contain other click-to-reveal menus and palettes that seem to multiply as the computer increases in power. The computer has an infinite amount of capacity to HIDE in order to create the illusion of simplicity. Now that computer screens are shrunken onto cell phones, microwave ovens, and every manner of consumer electronics, the power to HIDE incredible amounts of complexity is everywhere.

Hiding complexity through ingenious mechanical doors or tiny display screens is an overt form of deception. If the deceit feels less like malevolence, more like magic, then hidden complexities become more of a treat than a nuisance. The ear-catching "click" when opening a Motorola Razr cell phone or the cinematic performance of an on-screen visual in Apple's Mac OS X creates the satisfaction of owning the power to will complexity from simplicity. Thus complexity becomes a switch that the owner can choose to flip into action on their own terms, and not by their device's will.

SHRINK-ing an object lowers expectations, and the hiding of complexities allows the owner to manage the expectations himself. Technology creates the problem of complexity, but also affords new materials and methods for the design of our relationship with complexities that shall only continue to multiply. Although instilling "pity" and choosing how to "control" it sound like draconian approaches to simplicity, they can be seen in a positive light for the feelings of enjoyment they create.

SHE: EMBODY

As features go into hiding and products shrink, it becomes ever more necessary to embed the object with a sense of the value that is lost after HIDE and SHRINK. Consumers will only be drawn to the smaller, less functional product if they perceive it to be more valuable than a bigger version of the product with more features. Thus the perception of quality becomes a critical factor when making the choice of less over more.

EMBODY-ing quality is primarily a business decision, more than one of design or technology. The quality can be actual, as embodied by better materials and craftsmanship; or the quality can be perceived, as portrayed in a thoughtful marketing campaign. Exactly where to invest—real or believed quality—to get maximum return is a question with no single definitive answer.

Perceived excellence can be programmed into consumers with the power of marketing. When we see a super-athlete like Michael Jordan wearing Nikes, we can't help but imbue the sneakers with some of his heroic qualities. Even without the association of a celebrity, a marketing message can be a powerful tool to increase belief in quality. For instance, although I'm usually loyal to Google, I've been recently exposed to a bevy of Microsoft live.com and Ask.com television commercials and now I find myself Google-ing much less. The power of suggestion is powerful.

Embodying an object with properties of real quality is the basis of the luxury goods industry and is rooted in their use of precious materials and exquisite craftsmanship. Relatedly, a designer of Ferrari cars once told me that a Ferrari has fewer parts than a common car, but the parts themselves are significantly better than anything else on this earth. This elegant tale of construction uses the simple philosophy that if good parts can make a great product, incredible parts can lead to a legendary one. Sometimes there are instances of overkill, such as the titanium-clad laptop I own—I'm unlikely to need titanium's strength to shield myself from a bullet. But I enjoy the personal satisfaction that a higher quality material is used instead of an inferior plastic. The upside of materialism is that the way something we own feels can change how *we* feel.

Sometimes mixing actual and perceived qualities works well, like in the design of the Bang & Olufsen remote control. The unit is thin and slender in composition and made with the finest materials, but is significantly (and intentionally) heavier—as a means to subtly communicate higher quality—than you would expect from its appearance. Substantive technologies, like three CCD imaging arrays inside a video camera instead of the standard single array, are usually invisible. Thus the perception needs to be made visible somehow, unfortunately in direct contradiction to HIDE. An unobtrusive sticker on the unit like "3 CCD's" or a message that appears when the unit is first turned on helps to advertise this extra hidden power. It is necessary to advertise qualities that cannot be conveyed implicitly, especially when the message of embodiment simply tells the truth.

SHE SHE'D

Lessen what you can and conceal everything else without losing the sense of inherent value. EMBODY-ing a greater sense of quality through enhanced materials and other messaging cues is an important subtle counterbalance to SHRINK-ing and HIDE-ing the directly understood aspects of a product. Design, technology, and business work in concert to realize the final decisions that will lead to how much reduction in a product is tolerable, and how much quality it will embody in spite of its reduced state of being. Small is better when SHE'd.

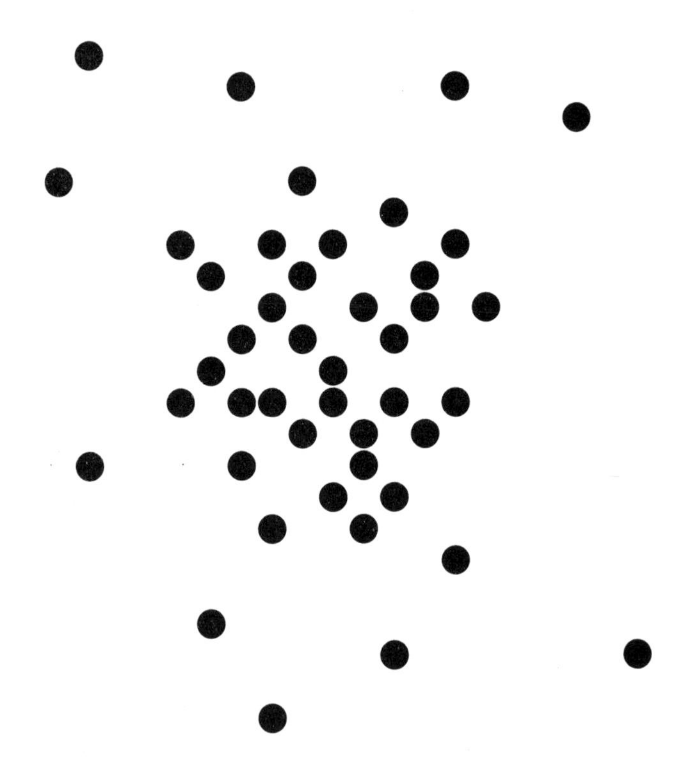

LAW 2

Law 2

ORGANIZE

Organization makes a system of many appear fewer.

The home is usually the first battleground that comes to mind when facing the daily challenge of managing complexity. Stuff just seems to multiply. There are three consistent strategies for achieving simplicity in the living realm: 1) buy a bigger house, 2) put everything you don't really need into storage, or 3) organize your existing assets in a systematic fashion.

These typical solutions have mixed results. At first, a larger home lowers the clutter to space ratio. But ultimately, the greater space enables more clutter. The storage route increases the amount of empty space, but it can be immediately filled in with more stuff that will need to go into storage. The final option of implementing a system takes the form of things like closet organizers, that help bring structure to the chaos as long as the organizing principles can be obeyed. I find it compelling that all three clutter-reducing industries—the real estate market, easy storage services such as Door to Door, and rational furnishing retailers like the Container Store—are booming.

Concealing the magnitude of clutter, either through spreading it out or hiding it, is an unnuanced approach that is guaranteed to work by the first Law of REDUCE. There are only two questions to ask in the de-complicating procedure: "What to hide?" and "Where to put it?" Without much thought and enough hands on deck, a messy room becomes free of clutter in no time, and remains so for at least a few days or a week.

However, in the long term an effective scheme for organization is necessary to achieve definitive success in taming complexity. In other words, the more challenging question of "What goes with what?" needs to be added to the list. For instance in a closet there can be groupings of like items such as neckties, shirts, slacks, jacket, socks, and shoes. A thousand-piece wardrobe can be organized into six categories, and be dealt with at the aggregate level and achieve greater manageability. *Organization makes a system of many appear fewer.* Of course this will only hold if the number of groups is significantly less than the number of items to be organized.

Working with fewer objects, concepts, and functions—and fewer corresponding buttons to press—makes life simpler when faced with the alternative of having too many choices. Nevertheless, making the right decisions to achieve integration across disparate elements can be a complex process that easily exceeds the trivial task of organizing one's closet. Here we look to describe the simplest ideas to help you get on your way.

SLIP: WHAT GOES WITH WHAT?

Matching up pairs of socks as they've just come out of the wash is easy when they are all the same make and model. Unfortunately most effects that come our way are not as simple as a generic pair of black stockings. Seeing the forest(s) from the trees is a common goal that is made easier by an ad hoc process I call SLIP: SORT, LABEL, INTEGRATE, PRIORITIZE.

SORT: Write down on small post-it notes each datum to be SLIP-ped. Move them around on a flat surface to find the natural groupings. For example, let me SLIP my own mind with the urgent and undone tasks for today: mit press, maharam, peter, kevin, amna, annie, burak, saéko, reebok, t&h, dwr, and so forth. Manually moving them around and placing them next to each other results in the rough groupings below.

LABEL: Each group deserves a relevant name. If a name cannot be decided upon, an arbitrary code can be assigned such as a letter, number, or color. Realizing the proficiency to SORT and LABEL requires practice like any major professional sport.

NOW	2ND YR	1ST YR	NOW+	NEW	NEAR	FAR
amna	annie	brent	wired	maharam	peter	saéko
mike	burak	isha	mit press	reebok	kevin	atsushi
	kelly	amber		t&h		seung-hun
	danny			dwr		lisbeth

INTEGRATE: Whenever possible, integrate groups that appear significantly like each other. Some groups will break apart at this phase. In general, the fewer the groups the better.

NOW	RESEARCH	NEW	NEAR
wired	annie, brent	maharam	peter, saéko
mit press	burak, isha	reebok	seung-hun, atsushi
amna	kelly, amber	t&h	lisbeth, kevin
mike	danny	dwr	

PRIORITIZE: Finally collect the highest priority items into a single set to ensure that they receive the most attention. The Pareto Principle is useful as a rule of thumb: assume that in any given bin of data, generally 80% can be managed at lower priority and 20% requires the highest level. Everything is important, but knowing where to start is the critical first step. The Pareto assumption makes it simple to focus on the "vital few."

FOCUS	BASE	NEXT
wired	annie, atsushi, danny, brent	maharam
mit press	burak, saéko, seung-hun, amber	reebok
amna	kelly, peter, lisbeth, isha	t&h
mike	kevin	dwr

As presented above, SLIP is a free-form process for finding answers to the question of "What goes with what?" The many little bits of cut-up post-it notes on my desk are the system of chaos brought to order with my fingertips. Finding the organizational scheme that works best for you is a wise investment.

There is no science to SLIP, so there isn't any right or wrong to the method. You should adapt it as you see fit. If you slip (sic) there is nobody to watch you fall, so it's worth a try. A computer tool for playing with the SLIP process is available for free on *lawsofsimplicity.com* in case you don't like little piles of paper sitting on your desk.

TAB(LES)

Getting organized is the theme of this Law, and SLIP is one of many ways to get you started. Googling "organization methods" will give you several million more varieties, like the popular "mind map" technique where related elements spider out radially like spokes on a wheel. In addition, a thorough search of the Web will reveal three-, and four-dimensional algorithms for organizing thoughts with accompanied visual acrobatics that astound. Animated text grows from trees, images pop out of a fishbone structuring pattern, and ideas float and fly in realistic 3D landscapes.

The visual presentation of information is a topic that I'm supposed to know a few things about as it represents a cornerstone of my career. Yet no matter how much I learn about the intricacies of graphic design, I always end up at the same place: the "tab" key. In the days of the typewriter, it was the tab key that could lend the magic possibility of creating order from chaos. The tradition of the tab key still lives on in the age of the word processor, but the satisfying *"thunk"* sound of the typewriter's advancing to a tab is unfortunately lost. Most undergraduate students return the quizzical look of "typewriter?"

The relevance of the tab key to the concept of organization is that it is the one key on the keyboard that is designed to make information simpler. Consider the following list of items:

red lion cola pepper sapphire
blue bear frappe salt diamond
green alligator martini msg topaz
pink flamingo espresso garlic ruby
white giraffe milk cumin emerald
black penguin beer saffron amethyst
gray dog water cinnamon turquoise

As posed, their system of conceptual organization is not clear. Complexity is remedied with a generous sprinkling of tabs, and then the categories come to life—order emerges.

red	lion	cola	pepper	sapphire
blue	bear	frappe	salt	diamond
green	alligator	martini	msg	topaz
pink	flamingo	espresso	garlic	ruby
white	giraffe	milk	cumin	emerald
black	penguin	beer	saffron	amethyst
gray	dog	water	cinnamon	turquoise

The tabular form of viewing data is by no means rocket science, but it is a rare sort of visual magic that always works. In the medium of text, tabs break up the linear space of a document such that the paragraphs can stand out as the organizing principle. Beyond the English language paradigm, computer programming codes are written in a special dialect that often suffers from legibility. It is the well-tabbed program codes that are known to be the sign of an enlightened mind. When used strategically, tabbing, and similarly the use of the space and return keys, gifts the chaos of clutter with the lightest touch of visual design.

"What program do you use?" is a question I often get about the slides I use to present my work. I have concluded that the proper answer to the question is to counter-suggest the asking of a different question, "What *principle* do you use?" The plain, unadorned horizontal and vertical gridding of information lacks sex appeal, but it is the one sure thing in the vocabulary of graphic design. Whenever I get confused, I turn my eye to the furthest left-hand side of the keyboard. The quick path to simplicity is only a pinkie away.

THE GESTALT OF THE IPOD

In both perceiving and visually representing the natural organization of objects, we are supported by the mind's powerful ability to detect and form patterns. With matters of the visual mind, the school of Gestalt psychology is particularly relevant. Gestalt psychologists believe that there are a variety of mechanisms inside the brain that lend to pattern-forming. For instance, when you see a box made with a single connected penstroke that is not completely closed, your mind can essentially "fill in the blank" and imagine it closed. Another example of Gestaltism is the tendency to mentally continue a series of drawn figures like "circle, circle, circle" with another circle.

Allow me to draw an illustration that helps to complete the gestalt of Gestalt psychology.

What's the difference between the cluster of 30 dots displayed on the left, and those on the right? The answer is simple. On the left there is no order to the randomly placed dots; on the right there is a clear grouping of some of the dots. We immediately pick out the group of dots as a "whole," even though it's composed of many little dots. In effect by gathering the dots into the group as on the right, we have simplified the otherwise haphazard display of 30 dots by giving order to the chaos.

Humans are organization animals. We can't help but to group and categorize what we see. Is he a poser? Is she a doll? Are they together or traveling separately? Does this top go with this bottom? The principles of Gestalt to seek the most appropriate conceptual "fit" are important not only for survival, but lie at the very heart of the discipline of design. Germany is arguably the country that originated the design field through its legendary Bauhaus school founded in 1919. Thus it is a little more than coincidence that the German word for design is *gestaltung*. Traditionally, German companies like BMW, Audi, and Braun have stood for design solutions that aspire to fit perfectly with the mind. Their common goal has been to relentlessly find the most appropriate gestalt that befits a need.

The changing gestalt of the Apple iPod reveals how small changes in organization create big differences in a design.

When it first came out, the controls were laid out as follows:

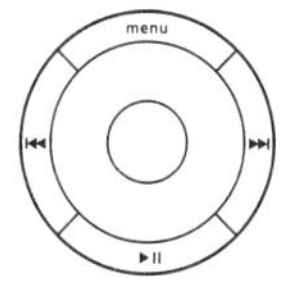

Then, perhaps as a cost reduction technique, or due to complaints from people with fat fingers, Apple separated the four buttons surrounding the jog dial into a discrete row of buttons in the subsequent version of the iPod:

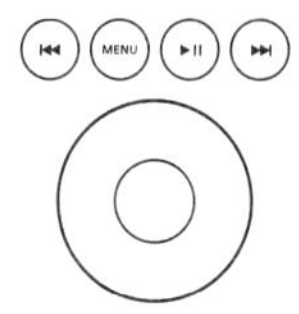

Apple had made the iPod more complex. Displacing the previously centralized functions to the unattractive row at the top made the newer iPod look complicated. I recall running out to buy one of the older iPods when this version with the button row came out. I was extremely irate because they had changed something from beautifully simple to unnecessarily complex.

In the newer versions, they have oscillated towards extreme simplicity by integrating all of the buttons into a single seamless control:

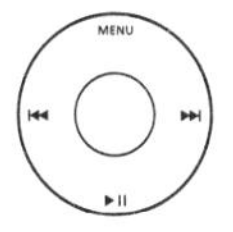

Let's look at all three designs placed side by side:

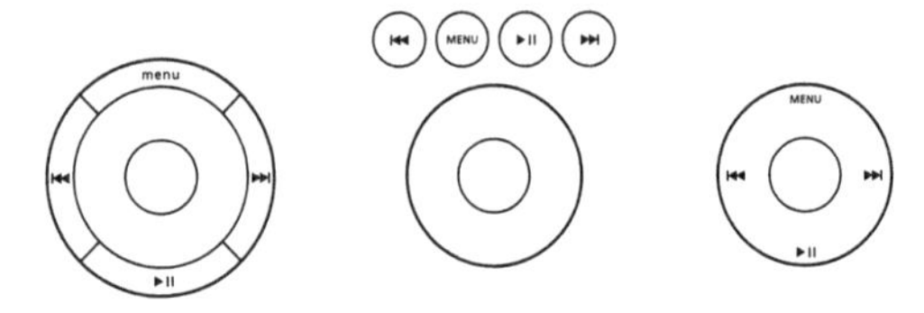

From left to right we can read this sequence of iPod evolutionary steps as "starting simple, then getting complex, and ending up as simple as possible." Translating the iPod controls into my dot diagrams looks something like this:

On the left the buttons are wrapped around the scroll dial, in the middle they are separated, and on the right they are integrated into a cloud where scroll dial and buttons are one. The right diagram of the cloud of dots represents where all of the individual elements have melted into one as if they were optically blurred through a lens.

The aesthetics of the blur are common in the history of art, ranging from the Impressionist paintings by Monet and his hazy clouds of tiny brushstrokes, to the stylized images of flowers by artist Georgia O'Keeffe. Soft-edged representations have an allure of mystique, and are thus inviting in nature. Similarly, the third phase of the iPod control is desirable because it blurs all controls into one image of simplicity.

There are downsides to the blurred approach, as evidenced by my dear brother-in-law's recently observed inability to operate an iPod for the first time at a Christmas party. It was not clear to him how to scroll through songs due to the integration of the buttons with the scroll dial. The question with which we began this journey, "What goes with what?" is answered by the blurred approach with simply, "Everything." I then remembered that everyone isn't necessarily a lover of abstract art and subjective interpretation. Everyone has their own gestalt, and that is why other MP3 players still sell. But eventually my brother-in-law did master the iPod to his glee, proving that the iPod control wheel *can* be a good gestalt.

SQUINT TO OPEN YOUR EYES

Groups are good; too many groups are bad because they counteract the goal of grouping in the first place. Blurred groupings are powerful because they can appear even more simple, but at the cost of becoming more abstract, less concrete. Hence simplicity can be a creative way of looking at the world that is driven by design. It feeds the mind's natural hunger to solve puzzles and to find the right gestalt.

The best designers in the world all squint when they look at something. They squint to see the forest from the trees—to find the right balance. Squint at the world. You will see more, by seeing less.

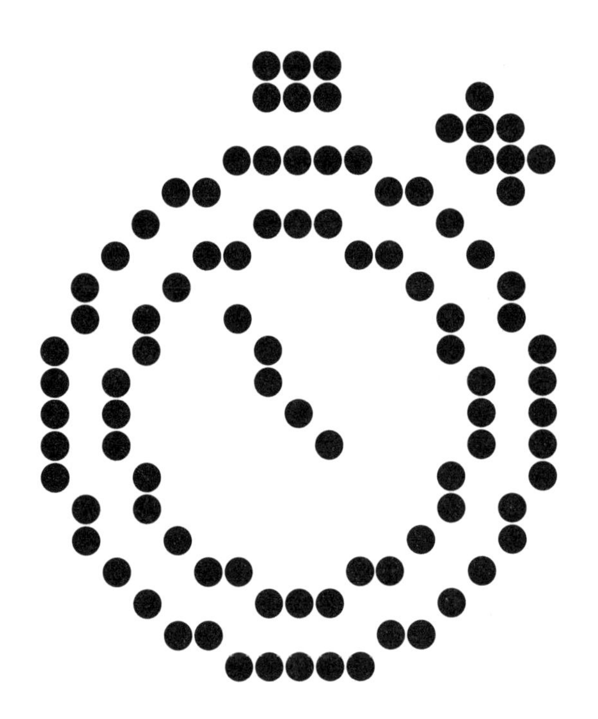

LAW 3

Law 3

TIME

Savings in time feel like simplicity.

The average person spends at least an hour a day waiting in line. Add to this the uncountable seconds, minutes, weeks spent waiting for something that might have no line at all.

Some of that waiting is subtle. We wait for water to come out of the faucet when we turn the knob. We wait for water on the stove to boil, and start to feel impatient. We wait for the seasons to change. Some of the waiting we do is less subtle, and can often be tense or annoying: waiting for a Web page to load, waiting in bumper-to-bumper traffic, or waiting for the results of a dreaded medical test.

No one likes to suffer the frustration of waiting. Thus all of us, consumers and companies alike, often try to find ways to beat the ticking hand of time. We go out of our way to find the quickest option or any other means to reduce our frustration. When any interaction with products or service providers happens quickly, we attribute this efficiency to the perceived simplicity of experience.

Achieving notable efficiencies in speed are exemplified by overnight delivery services like FedEx and even the ordering process for a McDonald's hamburger. When forced to wait, life seems unnecessarily complex. *Savings in time feel like simplicity.* And we are thankfully loyal when it happens, which is rare.

Then there's the implicit benefit: reducing the time spent waiting translates into time we can spend on something else. In the end it's about choosing how we spend the time we're given in life. Shaving ten minutes off of your commute home translates to ten more minutes with your loved ones. Thus a reduced wait is an invaluable reward not only with respect to business, but to life and your well-being.

Saving time is really about reducing time, and SHE as introduced in the first Law can help us. SHE says that we can realize the perception of reduction through shrinking and hiding, and can also make up for what is lost by embodying what is most important in subtle ways. Let's see if SHE is right again.

SHE: SHRINKING TIME

As a prototypical "busy guy" who's trying to stay sane, I'm personally familiar with the goal of shrinking time. I'm the guy who unties his shoes and removes his laptop from his bag before he reaches the table at airport security, in the hope of passing through with the speed of an Olympic downhill skier. Getting home before the kids are asleep is another daily challenge—one to which I apply sophisticated routing algorithms that get me from MIT to my house with the efficiency of a New York City messenger. In the former case I risk embarrassment while self-exposed in the security line, and in the latter case I up my premiums by swerving through the infamous battlefield of Boston traffic. My personal risks when saving time, however, are small compared to the larger scale at which businesses risk.

Reducing a five-minute task to one minute is the *raison d'etre* of operations management, the field that has brought us a world that never sleeps and is always on time. Superior operation management techniques played an important role in the rise of Toyota over GM in 2006. Promises of radio-frequency ID (RFID) tag technology that can uniquely identify every single product stocked on shelves will make taking inventory happen instantly. Businesses take great risks to optimize their processes out of the need for survival. At the individual level, we're also in the business of survival but we also have certain freedoms that allow us to play a different tune.

Of the infinite ways to whittle away at time, a superior solution is to remove all constraints, as I learned upon the introduction of Apple's iPod Shuffle. The Shuffle differs from other iPod products in that it has no display besides a single LED, and thus its user interface is vastly reduced at the gain of a lower price point and better resistance to wear.

I first heard about the Shuffle in a radio commercial that went something like, "Plug it in and get a completely random mix of your music library. That's right, completely random!" I couldn't contain my enthusiasm, and I began wondering: after Apple invented the usage of white in product design, had it now invented randomness?

Giving up the option of choice, and letting a machine choose for you, is a radical approach to shrinking the time we might spend otherwise fumbling with the iPod's scroll-wheel. The Shuffle's approach is to generate random choices, but we can foresee a future in which the iPod knows your preferences, habits, and even your moods and will play music accordingly. Eventually Google's "I'm feeling lucky" search option won't have to be lucky at all and will find the exact thing you're searching for.

A version of this future is already with us today. Go to Amazon.com and it recommends a handful of books you might like, based on the preferences of people it deems similar to you. Choosing to browse Amazon.com's entire inventory would be a time intensive task, and thus by caring less we can find savings in time. Letting someone else make the unimportant choices for us can be a sound coping strategy.

At a macroscopic level, governments and corporations go to great lengths to shrink time and cut corners as a means to reduce cost; at a personal level we make similar sacrifices that realize similar rewards in the name of efficiency. At the end of the day, there is an end of the day. Thus choosing when to care less versus when to care more lies at the heart of living an efficient but fulfilling daily life.

SHE: HIDING AND EMBODYING TIME

Shrinking the time of a process can sometimes only go so far, and so an alternative means to "saving" time is to hide its passage by simply removing time displays from the environment. I stopped wearing a wristwatch many years ago as I found, like many others, that as a result I never feel that I am running out of time. Although even without a wristwatch, my cell phone volunteers the current time. I wish I could turn the display off.

Few examples exceed the slippery trick that casino parlors in Las Vegas play on their guests. Walking into a professional casino for the first time can be a disorienting experience. Typically there are no clocks or even windows to reveal the general time of day. This simple environmental setup reinforces your impression that you might be logically awake enough to gamble. I would imagine that if it were legal, casinos would want to reprogram all cell phones in their vicinity to display time in a garbled fashion in order to keep you there. Of course, hiding time does not save time; it simply creates the illusion that time is not of pressing concern.

When we see the frozen hands of a clock with a dead battery, and we sit there and watch it, we tend to have a sinking feeling. Something feels wrong. We like to see time flow, as it is only natural that it seek its natural progression forward. On the other hand, when a clock is completely hidden we tend not to question its flow and instead experience an unsettling sense of uncertainty as to what time it might be. Seeing a clock's second-hand *tick-tick* forward is a reassuring sign that all is well.

In the early days of personal computers, the transfer of data from internal memory to an external storage medium such as a disk drive or remote computer could take anywhere from a few seconds to many hours. You would execute the transfer command and wait until it ended—not knowing how long it might take. A frozen computer is like a frozen clock, and thus ways to psychologically deal with this torturous experience of waiting emerged in the form of "progress bars." When Apple used to invest in research, they conducted an experiment in which a user was presented with a task that required significant processing time. They found that when a graphical display of progress, or a "progress bar," was shown, the user would perceive that the computer completed the task in less time than when no progress bar was shown at all.

Let's do an experiment, shall we? Below on the left is a progress bar that is displayed as consecutive frames in time. Read them top to bottom, and you see that at the very end, the bar is fully filled. On the right is a progress bar that shows progress forward in increments until it reaches its fully filled state in a step-by-step fashion.

What did you find? I'm convinced. Less time is felt to elapse in the progress bar on the right. On the left, time messily plops out like ketchup from a Heinz bottle; on the right, time is gently spread across a slice of bread like margarine with a butter knife.

Telling people how much time they have left to wait is a humane practice that is becoming more popular. Witness the increasing number of crosswalk signals that have their own progress bar or numerical countdown display to show the time that remains. When waiting on hold for a service representative, an automated voice tells you how many minutes you may have until you speak to a human. Time can be embodied in the face of a clock, in digital form, or in an abstract graphical display. There are cases when at the minimal level of display a simple LED blinks monotonously as a kind of visual heartbeat to signal to its audience that everything is okay. Knowledge is comfort, and comfort lies at the heart of simplicity.

Time can be embodied through a more deceptive approach—using "styling" to create the illusion of motion and speed. A designer in the 1930s named Raymond Loewy is credited with a styling concept called "streamlining." You may not know his name, but you probably know the Coca-Cola bottle that he designed many years ago (I refer to the classic single-serve glass bottle, and not the bulbous one-liter plastic container used today). Loewy is known for being influenced by the aesthetics of flight and jet propulsion, and for transferring the "style" (not function) of flight onto regular household objects. For instance, a vacuum cleaner or toaster could be made to look more swift and light by giving it the visual characteristics of an airplane. A car could be made to look faster by attaching fins that had no aerodynamic function. Computers today use many of the swoopy styling cues from the automotive industry to enhance the image of speed. Alienware, now a Dell subsidiary, leads this trend to apply "hotrod" styling to a computer in the form of aggressive air ducts and theatrical lighting.

Styling is a form of deception that, although misleading, can be a desirable attribute from a consumer perspective. We need all the positive reinforcement we can get in order to feel that we are moving forward. Don't we?

TICK TICK TICK

Every year something like this happens: I get stuck on an airport runway for 4 hours in the middle of a snowstorm, then stand in line for 3 more hours to determine my future flight's fate, then the next morning stand for 2 hours in a line to get through security in order to wait another 1 hour on the runway again. The realization that life is about waiting comes later in life. As a child, the idea of waiting is something foreign and simply intolerable. But waiting is what we do in the adult world. We do it all the time.

Sometimes the mundane experience of waiting can reach dramatic heights. Like when you are about to give a presentation to an audience of hundreds and you are copying a critical file over from a thumb drive to the presentation computer and everyone's waiting for you to start, and the progress bar lazily marches along ... and ... then ... it *stops*. And *waits*. It tests your faith in the machine and silently taunts you to press "Cancel." Hundreds of eyes are on you. Do you have the guts to restart the process? Can the wait experienced now be gambled against what might be an even longer wait later? Feeling lucky?

Making critical processes run faster is a fantastic benefit to humankind. However fast doesn't come cheap. The cost of sending a document via the USPS is 39 cents but to send it overnight is $14.40—making it close to 40 times more expensive to get into the fast lane. A direct flight will save time over one with connections but will cost significantly more. Add in the interminably rising cost of fuel, and expect to continue paying an extra premium for the privilege of acceleration.

Web technologies are an exception to this time/cost tradeoff. Google News breaks stories that emerged only "3 minutes ago," giving you a free front row seat to world events as they happen. *Saturday Night Live*'s boastful "Live from New York" introduction doesn't seem like such a big deal when live webcasts are possible from anywhere in the world. The speed of the Web sets our expectations to *now*.

When speeding-up a process is not an option, giving extra care to a customer makes the experience of waiting more tolerable. I appreciate the free cookies and other samples in line at the Whole Foods store during the Thanksgiving season as the checkout queue snakes across the entire store. Saving time is thus the tradeoff between the quantitatively fast versus the qualitatively fast:

Restated in the terminology of SHE, SHRINK the time constraints on one hand and HIDE or EMBODY the dimension of time on the other hand. Saving time or staying in step with the flow of time—whichever costs the least to implement—will usually win the day.

SHE helps us to manipulate our relationship with time in favorable ways. When time is saved—or appears to have been—the complex feels simpler. A shot from the doctor hurts less when it happens quickly, and even less when we know that the shot will save our lives. This latter phenomenon is addressed in the fourth Law of LEARN, so let's not linger but move along so you do not have to wait.

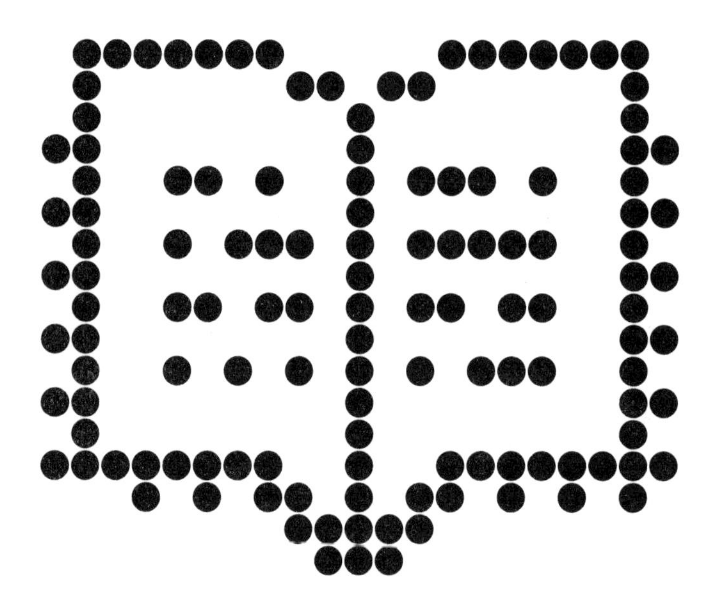

LAW 4

Law 4

LEARN

Knowledge makes everything simpler.

Operating a screw is deceptively simple. Just mate the grooves atop the screw's head to the appropriate tip—slotted or Phillips—of a screwdriver. What happens next is not as simple, as you may have noted while observing a child or a woefully sheltered adult turning the screwdriver in the wrong direction.

My children remember this rule through a mnemonic taught by my spouse, "righty tighty, lefty loosy." Personally I use the analogy of a clock, and map the clockwise motion of the hands to the positive penetration curve of the screw. Both methods are subject to a second layer of knowledge: knowing right versus left, or knowing what direction the hands of a clock turn. Thus operating a screw is not as simple as it appears. And it's such an apparently simple object!

So while the screw is a simple design, you need to know which way to turn it. *Knowledge makes everything simpler.* This is true for any object, no matter how difficult. The problem with taking time to learn a task is that you often feel you are wasting time, a violation of the third Law. We are well aware of the dive-in-head-first approach—"I don't need the instructions, let me just do it." But in fact this method often takes longer than following the directions in the manual.

Something as simple as teaching another person a basic concept might seem trivial in comparison to managing a complex supply chain or programming a supercomputer. However, anyone who has tried to teach a child the seemingly trivial task of tying shoelaces may suspect that writing code for Google's page-ranking algorithm is easier. As a professor at MIT, I freely admit that I'm still figuring out how to teach as I go. The single most helpful thing for my teaching was to experience the other side of learning: I became a student in an MBA program.

Becoming a student has allowed me to relive the humbling experience of being a freshman at MIT and feeling like the dumbest one on campus. Being a professor is the easiest thing in the world—you just have to act like you know all the answers. Being a student is much harder because you not only have to wring the answers from the cryptic professor, but you also have to make sense of them.

As a student and an educator, I present a few of my design-informed approaches to what I deem as "good learning." They represent a work-in-progress that patiently awaits refinement through the natural evolution of a living concept.

USE YOUR BRAIN

Learning occurs best when there is a desire to attain specific knowledge. Sometimes that need is edification, which is itself a noble goal. Although in the majority of cases, having some kind of palpable reward, whether a letter grade or a candy bar, is necessary to motivate most people. Whether there is an intrinsic motivation like pride or an extrinsic motivation like a free cruise to the Caribbean waiting at the very end, the journey one must take to reap the reward is better when made tolerable. However, reality TV shows like *Fear Factor* or *Survivor*—which I admit to having watched—prove that sometimes the reward alone justifies the journey regardless of how uncomfortable a path it might take.

The doctrine of "the carrot or the stick" points to a choice between positive and negative motivation—a reward versus a punishment. I disagree when teachers give their students candy and other perks for correct answers, but I also disagree with a colleague at MIT who throws erasers at students that fall asleep during class.

Instead, my ten years of data as a professor show that giving students a seemingly insurmountable challenge is the best motivator to learn. It is said that a massive amount of homework is a kind of reward for the average over-achieving MIT student. But after recently experiencing student life myself, I've lost my masochistic attitude in favor of a holistic approach:

BASICS are the beginning.
REPEAT yourself often.
AVOID creating desperation.
INSPIRE with examples.
NEVER forget to repeat yourself.

By now you've tired of my acronyms like SHE and SLIP so I won't tell you that the first letters of my mantra above spell BRAIN.

The first step in conveying the BASICS is to assume the position of the first-time learner. As the expert, playing this role is not impossible, but it is best ceded to a focus group or any other gathering of external participants. Observing what fails to make sense to the non-expert, and then following that trail successively to the very end of the knowledge chain is the critical path to success. Gathering these truths is worthwhile but can be time consuming or else done poorly. Hiring experts in the study of people, like anthropologists and human factors designers, is an effective method proven by the success of my friends at the international design consultancy IDEO. Then again, if you can't afford to retain IDEO and are willing to violate the third Law by taking a bit more TIME, the easiest way to learn the basics is to teach the basics yourself.

A few years ago, I visited the master of Swiss typographic design, Wolfgang Weingart, in Maine to give a lecture for his then regular summer course. I marveled at Weingart's ability to give the exact same introductory lecture each year. I thought to myself, "Doesn't he get bored?" Saying the same thing over and over had no value in my mind, and I honestly began to think less of the Master. Yet it was upon maybe the third visit that I realized how although Weingart was saying the exact same thing, he was saying it simpler each time he said it. Through focusing on the basics of basics, he was able to reduce everything that he knew to the concentrated essence of what he wished to convey. His unique example rekindled my excitement for teaching.

REPEAT-ing yourself can be embarrassing, especially if you are self-conscious—which most everyone is. But there's no need to feel ashamed, because repetition works and everyone does it, including the US President and other leaders. Simplicity and repetition are related, as supported by Slate.com's story on the re-election of George W. Bush in 2004 headlined: "Simplicity, simplicity, simplicity." On the campaign trail Bush delivered the same simple message on terrorism and Iraq repeatedly.

Artist Mike Nourse reinforced this point in his 2004 video artwork entitled, "Terror, Iraq, Weapons." Nourse started with Bush's televised speech on the eve of the invasion of Iraq and edited out all instances of three heavily repeated words: "terror," "weapons of mass destruction," and "Iraq." When Nourse spliced together just those clips, the resulting video amounted to ten percent of the speech. It's no surprise that the US subsequently went to war with Iraq, based upon a perception by many Americans that Iraq had weapons of mass destruction that were to be used in terror operations against the US. At the time I was certainly convinced and afraid like many others, and I wasn't sure why. Now I know. Repetition works.

AVOID-ing desperation is something to target when learning is concerned. We all want to "wow" people from the beginning with the newest bells and whistles in an amazing new product, but sometimes "wow" can become *"woah"* and you need an aspirin to cope with the anxiety of the overwhelming aspects of the new. I dread upgrading software on my computer because I know how eager the new program will be to tell me its latest and most wondrous features. The strategy of "shock and awe" can discourage the shocked-and-awed as I learned by experiencing the vast chasm of knowledge between teacher and learner as an MBA student. I also became aware of how professors can unknowingly become insensitive in a university setting. A gentle, inspired start is the best way to draw students, or even a new customer, into the immersive process of learning.

INSPIRATION is the ultimate catalyst for learning: internal motivation trumps external reward. Strong belief in someone, or else some greater power like God, helps to fuel belief in yourself and gives you direction. My own inspirational moment in design happened during my undergraduate years when I accidentally encountered a book by the eponymous designer and author Paul Rand. Rand's ubiquitous contributions to the landscape of American corporate icons, such as the logos for IBM, ABC, Westinghouse, and UPS, have provided aspirational goals for legions of designers. I met Rand at his studio exactly ten years after happening upon his book and forever treasure those memories. He died a year later at the age of 82, and the image I keep of him in my mind is his almost constant, loving embrace of his wife Marion. Rand taught me so much, in so little time.

Feeling safe (by avoiding desperation), feeling confident (by mastering the basics), and feeling instinctive (by conditioning through repetition) all satisfy rational needs. Inspiration from others serves a higher goal that, at least for me, is the true reward. The practice of education is the highest form of intellectual philanthropy.

Lastly, NEVER forget to repeat yourself. Might I have already said that?

RELATE-TRANSLATE-SURPRISE!

My five-step approach to the process of learning continues to evolve as an educator, but I began my career originally as an

MIT-trained engineer. During that period of my life, my peers taught me an important rule for learning complex systems: RTFM, short for "Read The F*cking Manual." Someone has a problem? Tell them, "RTFM." Case closed—the ultimate in simplicity. Of course, that solution isn't perfect. There may not be a manual available to read, for starters, and nobody really likes a potty mouth.

An alternative to the roughness of the "engineering approach" is the more sophisticated "designer approach" to easing the process of understanding. The best designers marry function with form to create intuitive experiences that we understand immediately—no lessons (or cursing) needed. Good design relies to some extent on the ability to instill a sense of instant familiarity. "Hey, I've seen this before!" is a targeted reaction that builds the confidence to give it a try. As you recall from the second Law, the Gestalt principles of design rely on our mind's ability to "fill in the blank" by synthesizing plausible relationships. Design starts by leveraging the human instinct to relate, followed by translating the relationship into a tangible object or service, and then ideally adding a little surprise at the end to make your audience's efforts worthwhile. Or writing these steps in shorthand: RELATE-TRANSLATE-SURPRISE!

The persistence of the desktop metaphor, introduced in the 80s, is a ubiquitous example of the impact of RELATE-TRANSLATE-SURPRISE. Prior to the graphical user-interface, the norm was a single, gridded screen large enough to display 80 by 24 characters of text. The entire world inside the computer was represented as a linear stream of digital alphanumeric codes. Researchers at Xerox leveraged the emergent graphics power of computers together with the common paradigm of an office desk to establish a recognizable relationship between a person and her information. Certain aspects of a physical desktop translated easily to the on-screen desktop: folders containing papers mapped to folders containing data files, and the physical waste basket mapped to a virtual trash can for deleted data.

The known relationship to a physical desktop forged immediate cognitive buy-in, which was reinforced by concepts that translated well. But there would need to be a substantial reward or otherwise meaningful "aha"-SURPRISE to warrant a switch to this so-called "disruptive" technology. That surprise manifested as the ability to collect, categorize, redistribute, and repurpose many more documents than previously imagined possible by moving to digital information management.

Successful cases like the "desktop metaphor" and other mappings between older customs and newer technologies have paved the way to make otherwise foreign experiences more familiar. RELATE-TRANSLATE-SURPRISE relies on having a common experience upon which to map your own, which unfortunately limits the approach to specific cultures and customs. For example, the original trash icon on the Apple Macintosh's desktop was unrecognizable to Japanese users who had never seen a vertically-ribbed metallic trash can. Metaphors serve to RELATE-TRANSLATE a key concept, but the SURPRISE can be undesirable when the metaphor doesn't work.

Design culture can also affect the way in which RELATE-TRANSLATE-SURPRISE operates. A more rational, typically German approach to design will diligently RELATE-TRANSLATE, but not necessarily warrant the SURPRISE ending. A Braun shaver works perfectly, period. Contemporary British design, on the other hand, can be characterized as being heavy on the SURPRISE factor as evidenced by Apple's innovative designs led by Briton Jonathan Ive. The intensely pleasurable quality of Italian design drives the inversion of RELATE-TRANSLATE-SURPRISE to SURPRISE-TRANSLATE-RELATE, such as Studio65's sofa inspired by a woman's lips. Thus there are as many ways of RELATE-TRANSLATE-SURPRISE as there are differing tastes.

Metaphors are useful platforms for transferring a large body of existing knowledge from one context to another with minimal, often imperceptible, effort on the part of the person crossing the conceptual bridge. But metaphors are only deeply engaging if they SURPRISE along some unexpected, positive dimension. For example the restaurants of chef Alain Ducasse are always throwing culinary curve balls—just when you think you know how something will taste, you discover unanticipated flavors. Great movies, like the films directed by M. Night Shyamalan, lull you into your comfort zone with identifiable plot elements such that everything makes perfect sense until the twist at the end. A metaphor used as a learning shortcut for a complex design is most effective when its execution is both relevant and delightfully unexpected.

THE REAL REWARD

Growing up, I found it odd that my classmates were rewarded with bicycles and cash incentives for getting good grades. When presenting this hard data to my parents their response was, "How lucky your friends are!" End of story.

Some reward systems stem from recognizing progress itself as the payoff. I witnessed this in my toddler as she grew from crawling on all fours to walking around like her older siblings. On the way from the kitchen to the dining room there is a single step down to a lower level. As she crawled from kitchen to dining room head first, she quickly learned the danger of this maneuver. Later she invented a way whereby she would turn her body around to let her legs down first, and successful navigation became possible.

When she began to walk, she attempted to go down the step with her not-yet-perfected walking process. She of course fell. I tried to show her that if she went down on all fours, she could use her previously devised method for navigating the obstacle safely. Unexpected to me, she refused to do so and wanted to walk down the step like everyone else. The reward, in this case, was growth. When we're older, we tend to forget this simple but key motivation we all had as children.

I find it odd that the cell phone I use is much smaller than the manuals that came with it. True, that which is difficult to use is proportionally difficult to learn. So a complex object warrants an equally complex instruction manual. But the manual that comes with my car is slimmer than the one for my digital camera. That's not a fair comparison, of course. To drive a car in the US, I must undergo formal instruction for a semester, log many hours of practice, and ultimately pass a licensing test. Thus taking "Driver's Education" in high school exempted me from needing a thicker instruction manual for my car.

Difficult tasks seem easier when they are "need to know" rather than "nice to know." A course in history, mathematics, or chemistry is nice to know for a teenager, but completing driver's education satisfies a fundamental need for autonomy. In the beginning of life we strive for independence, and at the end of life it is the same. At the core of the best rewards is this fundamental desire for freedom in thinking, living, and being. I've learned that the most successful product designs, whether simple, complex, rational, illogical, domestic, international, technophilic, or technophobic, are the ones that connect deeply to the greater context of learning and life.

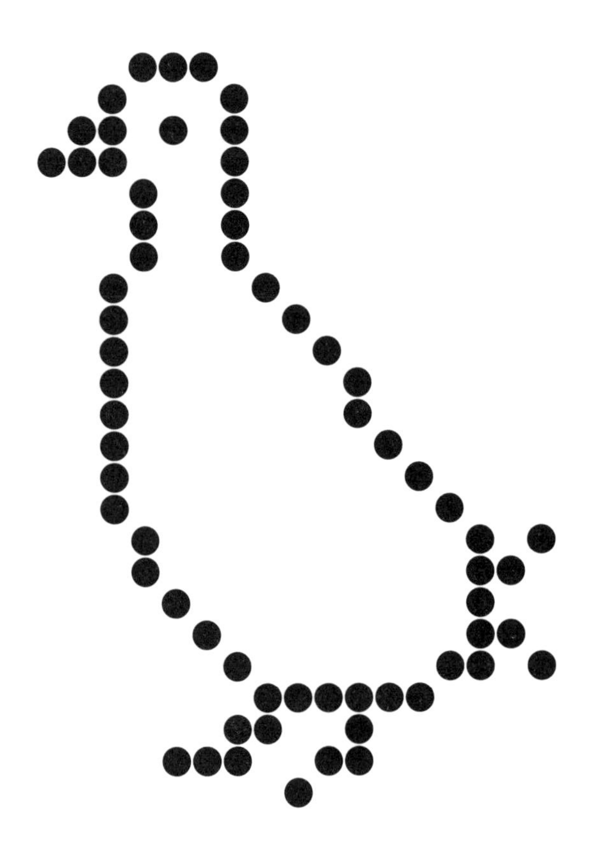

LAW 5

Law 5

DIFFERENCES

Simplicity and complexity need each other.

Nobody wants to eat only dessert. Even a child that is allowed to eat ice cream three meals a day will eventually tire his sweet tooth. By the same token, nobody wants to have only simplicity. Without the counterpoint of complexity, we could not recognize simplicity when we see it. Our eyes and senses thrive, and sometimes recoil, whenever we experience differences.

Acknowledging contrast helps to identify qualities that we desire—which are often subject to change. I don't personally prefer the color pink, but I do like it as a dash of brightness in a drab sea of olive green. The pink appears bold and vibrant as compared with its dark and muted surroundings. We know how to appreciate something better when we can compare it to something else.

Simplicity and complexity need each other. The more complexity there is in the market, the more that something simpler stands out. And because technology will only continue to grow in complexity, there is a clear economic benefit to adopting a strategy of simplicity that will help set your product apart. That said, establishing a feeling of simplicity in design requires making complexity consciously available in some explicit form. This relationship can be manifest in either the same object or experience, or in contrast with other offerings in the same category—like the simplicity of the iPod in comparison to its more complex competitors in the MP3 player market.

Within the same experience, finding the right balance between simplicity and complexity is difficult. Achieving a situation where the differences enhance, instead of cancel out, either's existence is something of a subtle art that I am still unclear about. The closest approximation to a solution I have found is in the concept of rhythm, which is grounded in the modulation of difference.

Think of a mathematical graph going upwards to complexity, then sloping downward to simplicity, then upward to complexity, and back down again ad infinitum. You can think of this happening over time, like a song that changes throughout its development; or else you can think of it as happening in space, like a painting where your eyes travel across the image and the experience changes. The rhythm of how simplicity and complexity occur in time and space holds the key.

NO RHYTHM AT ALL

In the age of electronic networking with services like LinkedIn and Friendster, the practice of sharing business cards is gradually losing its value. Nonetheless, having been raised in the business culture of Japan, where the exchange of cards is a formal act, I am still attached to the custom of presenting my business card held between thumb and forefingers of both hands while politely bowing. In my early days there, I recall being scolded countless times by superiors for not carrying my cards with me. To present yourself to a stranger without a card was considered the utmost insult.

Times have changed in Japan, and the custom of the two-handed offering is giving way to globalism's more informal one-handed pass. The printing quality and craftsmanship of business cards has declined along with their importance. The phrase "Google me" seems to mark the eventual demise of the fine tradition of business carding.

Still the business cards seem to flow towards me in their regular shape of a rectangle generally measuring 2 inches by 3.5 inches in the US, or in Asia and Europe 55 millimeters by 90 millimeters. My desk is generally kept clear and organized according to the second Law. Thus when business cards begin to collect on my desk, action is necessitated. The pile of cards is sorted according to SLIP, entered into my database, and then proceeds directly to the recycling bin (assuming the cards are made of paper and not metal or plastic as they sometimes are).

In the interest of full disclosure I must admit that I have violated the second Law of ORGANIZE. There is one business card that has never made it to the waste dump. It is a thin, creme-colored card with an illustration of a mystical sheep. At first I attributed my inability to throw it out to the sheep's watchful look. Sometimes business cards are printed with pho-

tographs of the person, and I have no problem seeing these shredded so my reluctance to toss this card isn't caused by the presence of a witness. I do not know the person well—I met Hiroaki only once—so it holds no particular sentimental value. Yet the card has quietly sat on my desk for more than seven years now and is likely to remain.

Place your business card next to his card. The monochrome printing of this book does not convey the soft yellow of its paper stock, or the red highlight at the lower left corner in his illustrator's mark. But your mind can fill in the details. It remains on my desk because I have encountered nothing similar to it in size or pictorial character. It is the one business card that is not like the others. If thin business cards with pictures of farm animals became trendy, it would certainly lose its value.

TEA WITH TANAKA

I had the privilege of knowing the father of modern Japanese graphic design—Ikko Tanaka. (His first name in kanji means simply "one light.") Once while living in Japan, I attended a private tea party at Tanaka's residence together with the famed contemporary architect Shigeru Ban. The words "tea party" conjure up an image of finely woven doilies and petit fours, but the Japanese tea party is something simply sublime.

Tanaka had been a practicing student of *chanoyu* (chaw-noh-yoo), the tea ceremony, and we were his test subjects. It's hard to imagine someone so masterful still being a student in his 70s, but in Asia there are many examples of this continuous cycle of learning. In the martial art of Karate, for instance, the symbol of pride for a black belt is to wear it long enough such that the die fades to white as to symbolize returning to the beginner state. Tanaka was the black belt of Japanese design.

The ceremony began, as is customary in some styles of *chanoyu,* with an examination of the tea-making implements. We passed around tea "cups" (more like deep bowls) to admire. If I remember correctly, I was assigned the cup from the 18th century that looked something like a horrible accident at the kiln. It was a deep and shiny black ceramic bowl where all of its external surfaces seemed to wrap unintuitively in the manner of a Salvador Dali painting. It was far from clear where I should place my lips to the bowl.

There I was at the house of Japan's foremost master of Modernism sipping from something completely imperfect, of non-platonic geometry (no cylinders, spheres, cubes to be found), that lacked all recognizable features of a cup. Visibly it was completely imperfect—lacking the smooth and white surfaces of simplicity as commonly sold today in the dishware section of the Ikea shop.

For this reason, however, the other elements of Tanaka's tools of tea came into view as pure perfection. Such as the lacquer tea-powder container from the 17th century where its matte black lid fit with its mate with the impossible precision of Lego blocks. Or else the subtle details of the wooden surfaces in his tea room that belied a lineage of tree that was nonexistent. The cup came to indirectly symbolize for me the essence of Japanese aesthetics, which strives for ultimate perfection. Its unexpected complexity made everything already impossibly simple, become even simpler.

FEEL THE BEAT

Taa taa ti ti taa. This is not some foreign language, but is the phonetic phrasing of rhythm that I learned from my music teacher in elementary school. *Ti ti ti ti taa taa. Rest. Ti taa ti taa ti ti ti ti taa.* It's all coming back to me. Hearing the counterpoint between long sound, short sound, and the absence of sound in the kind of sequence a jazz drummer can create engages the entire body in dance. On the other hand, if you create a simplistic rhythm like *"taa taa taa taa taa taa taa taa taa"* where the *taa*'s go on forever to sound out a monotonic beat, your audience will not bother to hang around for the last *taa.*

Consider in one day, the sequence of events to occur in the following pattern. *Complexity, complexity, complexity, complexity, complexity, complexity, complexity, complexity, complexity, complexity, complexity, simplicity.* Simplicity becomes salvation.

Simplicity, simplicity, simplicity, complexity, simplicity, simplicity, complexity, complexity, simplicity, complexity, complexity, simplicity, simplicity, complexity. It is the rhythm of simple and complex that matters the most.

Simplicity, simplicity, simplicity, simplicity, simplicity, simplicity, simplicity, simplicity, simplicity, simplicity, simplicity, simplicity, simplicity, simplicity, simplicity, simplicity, simplicity,

simplicity, simplicity, simplicity. There is no way to connect with simplicity when how complexity feels has been forgotten.

Alternatively, in the spatial domain consider a large canvas painted completely black versus another large canvas completely covered with scattered paint drippings like a bad Jackson Pollock interpretation. Both are monotonous expressions of simplicity and complexity in their distinctly separate forms. At the risk of sounding boring, I would place either painting on my wall at home for at least a day because I like to keep an open mind. Perhaps a simple dose of imagination applied to one of the pieces might prolong my attention span. For instance, a single image where parts are thoughtfully painted flat black, and other parts that are detailed with splatter would likely keep my interest for much longer. Variety tends to keep our attention when the rhythm of difference captivates.

There are some rhythms that we welcome their monotony, such as the changing of the seasons from winter, spring, summer, fall, and back to winter again.

Crunch, crunch, crunch. I recall walking through the snow in the middle of night through my quiet neighborhood only to hear my own breathing and footsteps. I reflected upon the fact that the snow of winter would eventually cease and give way to the green of spring. The combination of a silent night and my eventual advancement to middle age forced the rhetorical question, "How many more years might I experience a peaceful winter evening like this?" I am now more careful to feel the precious rhythm of each year of life. I hear the beat of simplicity and complexity quite clearly in everything that I experience. Can you hear it too?

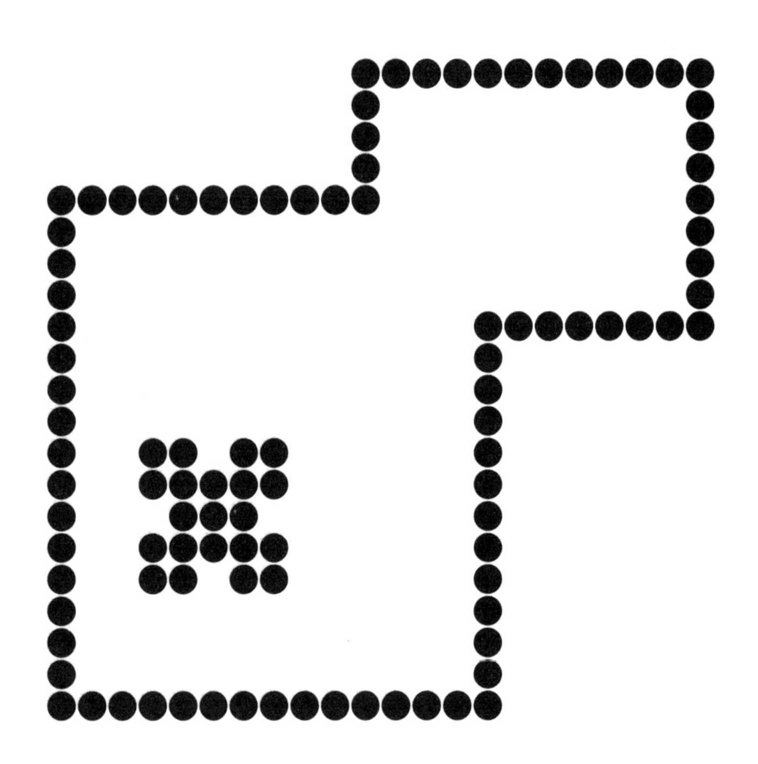

LAW 6

Law 6

CONTEXT

What lies in the periphery of simplicity is definitely not peripheral.

There is something about how our eyes and hands work in concert. Picture yourself at the pottery wheel, sculpting each detail with intense concentration. Everything that matters is happening in the foreground, at your fingertips, and is completely within your immediate field of vision. Your cell phone goes off or the doorbell rings, and this tightest of control loops is disrupted while the background surges into the foreground. Thankfully you notice that a pot on the stove is boiling over, or realize that your hand has been cut and unattendedly bleeds.

While the words "narrowness" and "focus" mean essentially the same thing, the former has a negative connotation while the latter has a positive one. An athlete who reaches the Olympics, for instance, is not "narrow" but focused. But focus isn't always a good thing.

I was once advised by my teacher Nicholas Negroponte to become a light bulb instead of a laser beam, at an age and time in my career when I was all focus. His point was that you can either brighten a single point with laser precision, or else use the same light to illuminate everything around you. Striving for excellence usually entails the sacrifice of everything in the background for the sake of attending to the all-important foreground. I took Negroponte's challenge as a greater goal of finding the meaning of everything around, instead of just what I directly faced.

What lies in the periphery of simplicity is definitely not peripheral. The sixth Law emphasizes the importance of what might become lost during the design process. That which appears to be of immediate relevance may not be nearly as important compared to everything else around. Our goal is to achieve a kind of enlightened shallowness. It is befitting that we start this trek by talking about nothing.

NOTHING IS SOMETHING

Science holds that entropy in the universe is always increasing. What does this mean in lay terms? A child opens an illustrated story book, flips through the pictures, and sees an empty part of the page. A crayon clenched in her fist, she moves her hand towards the blank space. What is she likely to do? Fill in the emptiness, of course.

This is the eighth book that I've designed and written, but it is the first that I have written more than designed. All designs have upheld the common priority of maximizing "white space"—essentially all these blank areas of the page that surround the text. Such surfaces invite chaos, in the same way that a countertop at home collects change, mail, keys, and so forth. Similarly we might scribble notes in these empty spaces that surround, and also in the gutters that separate lines of text.

Consider the simple challenge of a page that says nothing more than, "Don't write on this page." Can you resist the urge? Turn to page 57 and take the test.

The inviting, open space of the page challenges your sense of pride that five written words can hold you in command. Your natural inclination is to ask, "Why not?" With no explanation provided, we have to fill in the blank ourselves either literally with our pencil marks or else metaphysically with our own conclusions. Maybe it's due to the author's religion? Or perhaps it's a radical measure to conserve the global supply of ink? Sometimes we can be off target, but by this sixth Law of CONTEXT that means that we are actually *on* mark.

While visiting a shrine in Japan, I noted a large rectangular area that was carefully cordoned off by a rope decorated with white paper markers. The rectangle was empty, and carried an air of nobility because it was on grounds immediately nearby a temple. Could this be a sacred burial ground? I stood for many minutes contemplating the meaning of the emptiness, slipping into the same calm trance I had experienced in the adjacent Zen-style rock garden. A priest approached the mysterious rectangular zone, and waved to a car entering the temple grounds. The rope was untied and the car slipped into the space to receive its annual blessing to ward against accident and injuries. It reminded me that you don't have to be a Zen monk to appreciate empty space—especially if you're trying to park on a crowded street in Manhattan.

If given an empty space or any extra room, technologists would invent something to fill the expanse; similarly, business people would not want to pass up a potential lost opportunity.

On the other hand, a designer would choose to do their best to preserve the emptiness because of their perspective that *nothing* is an important *something*. The opportunity lost by increasing the amount of blank space is gained back with enhanced attention on what remains. More white space means that less information is presented. In turn, proportionately more attention shall be paid to that which is made less available. When there is less, we appreciate everything much more.

AMBIENCE IS EVERYWHERE

Look up from the book for a moment and glance around. What do you see? I see other tired passengers in the cramped space in which I type this passage on my little laptop computer. The sound of the engines is so loud that it is difficult to hear anything besides white noise. And the height of the seats prevents me from seeing more than the baldness of the passenger in front of me. The experience of riding an airplane can be one of uncomfortable isolation in almost all of the senses. Where there is so little of significance to feel, every minor sensation seems annoyingly amplified.

For instance, I try to dampen the ambient noise by wearing industrial ear plugs. Yet instead of silence, I hear the slow release of breath from my lungs. I wear a mask to cut out the overhead lights, yet the cloth of the mask chafes against my face, reminding me of its presence and its intended purpose. Small things in the environment matter more when you are forced to pay attention to them. Thus the background, or "ambient" environment, will take precedence over the fore-

DON'T WRITE ON THIS PAGE.

ground, or focused task, when there is nothing to fixate upon except everything that surrounds.

When going on a tropical vacation for pure relaxation purposes, embracing the ambience of the destination gives you necessary repose. The sum total of the many small details of the experience—the cleaner air, the higher percentage of smiles, the delicious tastes, and so forth—all add up to what is special. The hotel industry and other experience-based businesses require exhaustive attention to many minutiae that normally go overlooked at the individual level, but cumulatively achieve real relevance.

I once met a designer friend in a quiet Paris flat with white walls, white surfaces, and white furniture. A lunch of aesthetically prepared sushi was served. Red tuna, pink salmon, white squid, silvery mackerel, and a sliver of green leaf boldly engaged my visual senses as I took the entire scene into my mind. I reached to my chopsticks to begin, when my friend said, "The taste of this meal is affected by the room we sit in." True. With everything around me in pure white including the plate upon which the sushi was served, the thin slabs of raw fish atop the fist-sized mass of white rice appeared to float in space. I could imagine the taste to be very different in an environment that was appointed with different dishes, table, overall decorum, and even different people. Ambience is the proverbial "secret sauce" to any great meal or memorable interaction.

Creating white space—or, translating that to a room, "clean space"—enables the foreground to stand out from the background. However, the reality is that in everyday life we are unlikely to clear everything out with the ease of hitting the "delete" key on the word processor. The "taste" of any activity we face might be mixed in with the distaste of the clutter of our desk. But coincidentally the uplifting smile of a nearby child can sometimes help us to tune out any messes at hand. Being attuned to what surrounds us in the ambient environment can sometimes help us manage what's immediately in front of us. Synthesizing the ambient experience of simplicity requires attention to everything that seemingly does not matter.

COMFORTABLY LOST

In 2005 Google launched a service that allowed you to enter your address and see an overhead satellite map of your local vicinity. "There I am!" is the immediate impression, followed by, "There's everything else!" because you see all the houses and roads that surround you. Although you usually don't need a map to inform your location while sitting at home, there's a certain sense of comfort knowing that you can see the spot you occupy right there on the map. Interest in that web page diminishes once you have verified your location. The sense of comfort gives way to monotony.

Starting a book is easy, but somewhere in the middle you can be unsure of how far you are. A simple progress bar, with an X to mark the spot, can tell you exactly how far you've gotten, and how much more you have to go. Digital books require such displays, but for printed books like the one you hold in your hands, a quick squeeze on both left- and right-hand sides can provide your general locale. Page numbers and other traditional navigational elements like chapter headings are another layer of information that helps prevent you from getting lost. A progress bar printed on each page of this book, although favorably kitsch, would be overkill.

There is an important tradeoff between being completely lost in the unknown and completely found in the familiar. Too familiar can have the positive aspect of making complete sense, which to some can seem boring; too unknown can have the negative connotations of danger, which to some can seem a thrill. Thus there is a tradeoff between being found versus lost:

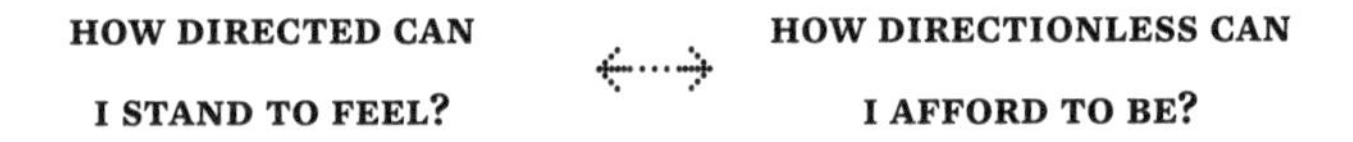

Your feeling of youth, state of health, and sense of adventure will dictate your preference for safety versus excitement to find the right balance where you can become “comfortably lost.”

I personally experienced this sensation of being “comfortably lost” on a recent vacation hike in Maine. I noted that the trails were marked with rectangles of bright blue paint. Each of the trails was highly navigable due to its good condition, but once in a while I would pause and wonder, “Where do I go next?” And almost like magic one of these blue markers that previously sat in the background of my perceptual field literally “popped” into the foreground. With my bearings restored, I would slowly return to the beautiful, uninterrupted forest vistas with the emotional satisfaction and comfort that one feels on a mountain hike.

If the forest were covered with ten times the number of blue markers I had seen on my hike, the probability of my getting lost would certainly be reduced. One could imagine the markers organized in some more symbolic shape—say a real arrow, instead of a cryptic linear marker. And if we wish to go that far, why not just paint the more explicit text, "This way," on the rocks in 100-point Helvetica so there's no ambiguity whatsoever? Yet at some point, with the successive addition of more sophisticated elements, the true value of the untainted forest suddenly vanishes.

The bridging experience that connects the foreground and background contexts can be made explicit as in a map, or less explicit as in the blue painted markers of the forest. Ample incorporation of empty space removes the need for a specific bridge between foreground and background because the navigation is implicit—you *can't* get lost.

Complexity implies the feeling of being lost; simplicity implies the feeling of being found. By the fifth Law of DIFFERENCES, transitions from simple to complex are a key consideration in the rhythm of feeling. In this sixth Law, we ask what happens between the beats, and question where you might be in the progress of the song. Once you have properly situated yourself, you're completely free to get lost in the rhythm.

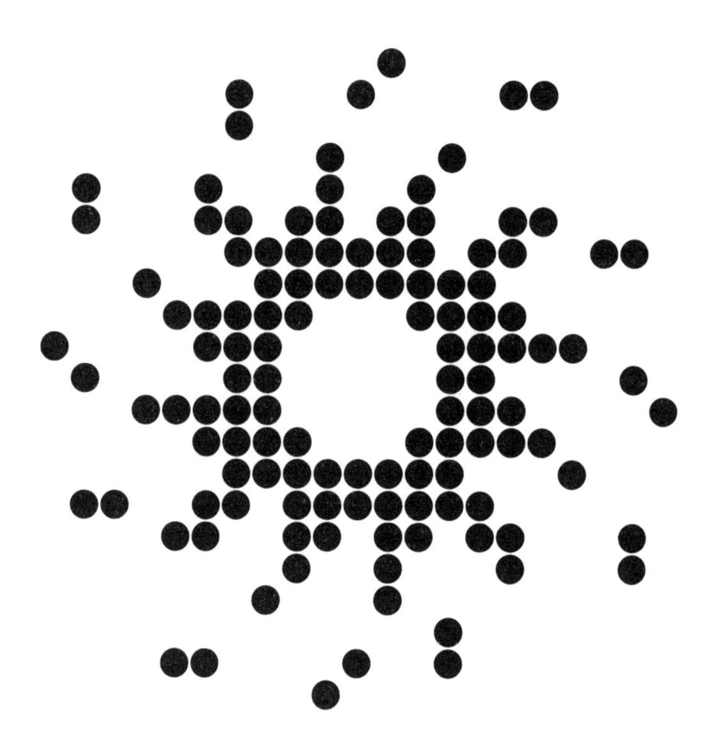

LAW 7

Law 7

EMOTION

More emotions are better than less.

Simplicity can be considered ugly. Take my mother who absolutely despises anything of neutral color or minimalist form. She wants neon flowers, bejeweled frogs, and other decorative essentials. When it comes to aesthetics, she's all about the "bling."

From a rational perspective, simplicity makes good economic sense. Simple objects are easier and less expensive to produce, and those savings can be translated directly to the consumer with desirable low prices. As evidenced by the extremely affordable line of simple products from furniture retailer Ikea, simplicity benefits the frugal shopper. However, there are some people, like my mother, who would say that simplicity is not only cheap, but would add that it *looks* cheap as well. A strong sense of self expression belies all of us humans, and many such decisions we make are not driven by logic alone.

The seventh Law is not for everyone—there will always be the die-hard Modernists who refuse any object that is not white or black, or else with clear or mirrored surfaces. My mother finds the iPod entirely unattractive. And while the older generation isn't Apple's targeted market (for the moment, at least), I am still the dutiful son I was raised to be, and so I find the seventh Law a necessary component in the simplicity toolbox. *More emotions are better than less.* When emotions are considered above everything else, don't be afraid to add more ornament or layers of meaning.

I realize this seems to contradict the first Law of REDUCE. But I use a specific principle to determine just the right kind of more: "feel, and feel for." Everything starts from being sensitive to your own feelings. Do you know how you feel? Right now? By connecting with the emotional intelligence inside yourself, the next step is to empathize with the environment that surrounds you. "Form follows function" gives way to the more emotion-led approach to design: "Feeling follows form." In this section we talk about emotion and the move towards complexity (and away from simplicity) that it sometimes requires.

FEEL, AND FEEL FOR: E-TIQUETTE

I've been emailing since 1984 when I arrived at MIT as a freshman. Although some fellow classmates had experience using Compuserve, the predecessor to online service companies like AOL, the concept of the network seemed quite foreign to me. I soon realized that everyone that mattered then had this odd device called a "modem" to connect to the computer network. So I got one and quickly became enslaved. I would check my email not just as a habit, but in lieu of breathing—my unhealthy fixation still haunts me. Which reminds me ... *There.* That one deep breath I just took will take care of the rest of the day ;-).

The smiley at the end of the paragraph causes the familiar tilt of the head to the left, and reveals a light touch of visual emotion. The Internet tells me that the smiley may have been invented in 1982 by a Mr. Scott Fahlman, currently at Carnegie Mellon University. I find it odd that in the long history of typeset text going back to Gutenberg that this invention had not happened sooner. The act of writing by hand doesn't lend itself to the use of smileys, however in the age of the typewritten letter, one would have expected to stumble upon the funny combination of characters that can make a wide variety of silly faces like :-) 8^) ;-o =) |-D and so forth.

Why have smileys evolved? Why does the textual medium need such baroque flourishes? Because of the human need to better express emotion—to capture the nuances of communication that we take for granted in speech. Interfacing through text, speaking to other disembodied voices, it is easy to stray from normal social mores. Smileys evolved as a way to temper and soften textual conversations without the facial cues speakers use to signify when they are "just kidding." And although sending photos is now possible, text continues to dominate.

My daughters send me email with text of all sizes, all colors, and sometimes in ALL CAPS! Not only does this seem to make their job of typing of email unnecessarily complex, it hurts my eyes! However I wholeheartedly accept their high-fidelity messages as I know their youthful exuberance cannot be contained by simple text messages alone. Does not the phrase "I love you!" have so much more meaning when typed, "I LOVE YOU!"? Think of it typed at 36 points in pink and bright yellow and it really can go over the top.

Much is said about the development from child to adult as a gradual process of neutering emotional output. Having the privilege of fostering minds and developing young careers on a daily basis, I can see evidence of people pressing the mute button on emotion every day. I once asked one of my students at MIT why she never smiled when communicating with others. She said, “Because I don’t want to look unprofessional.”

This event caused me to reflect on my own attempts to project professionalism as a professor, which caused a natural lean towards the stereotypical stern and authoritative. As an artist, I found the results of my self-analysis offensive. Thus, today I try to reply back to my daughters in all-caps and colorful letters when nobody’s looking, “I LOVE YOU TOO!!!”

FEEL, AND FEEL FOR: NUDE ELECTRONICS

When I first started a blog at MIT, I discovered that the most frequently accessed entry was the one entitled “nude electronics.” I could imagine the disappointment that a thrill-seeking geek might have had with my fully dressed prose.

By “nude electronics” I refer to the trend of making handheld consumer electronic objects smooth, seamless, and small to satisfy the market’s demand for simplicity. Using methods such as SHE, designers can simplify an object to its core and spare mysteriousness. But like a sheep that has been fleeced, you can’t help wonder if SHE is responsible for making the skinny little objects feel a tad bit cold.

The booming market for protective and decorative iPod accessories solves this problem—but it also raises a peculiar question. Why, after people are drawn to the simplicity of a device, do they rush to accessorize it? Why, as I browse the airport gadget store while waiting for a flight, do I see so many businessmen perusing Treo cases made of metal, plastic, leather, and cloth with the intensity of my younger daughters' choosing outfits for their Barbies?

Carrying cases for the simplicity object achieve two important goals. First of all, while SHE can make an object smaller, thereby alleviating the natural fear associated with larger and more complex machines, the successful application of SHE can instill a different kind of fear: concern for the object's survival. For instance, a student of mine is afraid to carry around his ultra-slim iPod Nano for fear of snapping it in half by accident. An iPod case provides needed protection for the pitifully undernourished and gaunt device.

The second reason is rooted in self-expression and in the need to balance the subzero coolness of the ideal consumer electronics gadget with a sense of human warmth. While the core object retains its pure, simple, and cool nakedness; its clothing can keep it warm, vivacious, and simply outrageous if so intended. The combination of a simple object together with a host of optional accessories gives consumers the benefit of expressing their feelings and feelings for their objects.

FEEL, AND FEEL FOR: AICHAKU

Growing up, my siblings and I were taught that everything in our environment, including inanimate objects, had a living spirit that deserved respect. "Even a cup?" we asked. "Even a desk?"

"Even a chewing gum wrapper?" "Even the house we live in?" The answer was always, "Yes."

Under this strict code of life, my taking a clean sheet of paper, crumpling it up, and throwing it away was grounds for punishment. I would be denying the paper's existence to perform a useful task, and divine retribution would result from the disrespect I had shown the paper. My family's belief system was based upon an extreme form of Shintoism, which is the ancient Japanese tradition of animism.

Believing that all things around you—rocks, river, mountain, and clouds—are somehow "alive" was something that I couldn't grasp as a child. However as an adult, I prefer the world with its mysteries intact and I find myself comfortable with the thought. In many animated works from Japan, like the work of acclaimed animator Hayao Miyazaki, the belief in the spirit living within all objects is, pun intended, alive and well.

Technology has helped to extend the illusion of life in a literal sense with robots that walk, talk, and even dance. Sony's AIBO robotic dog is constructed of plastic, motors, and a sophisticated computer. It obviously isn't a living dog, yet some AIBO owners relate to it almost as a real pet—gently stroking and coo-ing to them as if to express love for an animate, but non-living consumer product.

The Tamagocchi craze of the late 1990s also showed that anyone could fall in love with a small electronic keychain unit that yearned for human attention. Our yearning to care for what is purely imaginary extends to Neopets on the Web today where millions of cartoon characters are bred, fed, and loved. Although it contradicts traditional predominant Western religious beliefs, a kind of digital animism appears to be an acceptable and growing practice among our technologically empowered youths. If one can love an on-screen monster or a digital baby encapsulated in a little electronic box, is it so far a stretch to love and respect a plain piece of paper?

Modernism is the design movement that led to the clean, industrial look of many objects in our environment. It rejected unnecessary ornament in favor of exposing an object's truth through the raw materials of its production. Japan's rich tradition of almost perfectly crafted artifacts of wood and clay seems built on the same design principles as Modernism. However a hidden facet of Japanese design is this animistic theme. The precise lacquered surfaces of a bento box are more than just a fact of fine production; these surfaces—and the bento box that they comprise—are essentially alive. The inanimate box is accorded its own spiritual existence. There can be a natural emotional attachment to the object's life force that is a kind of deep, hidden ornamentation known to only those who feel it.

AI (love) **CHAKU** (fit)

Aichaku (ahy-chaw-koo) is the Japanese term for the sense of attachment one can feel for an artifact. When written by its two kanji characters, you can see that the first character means "love" and the second one means "fit." "Love-fit" describes a deeper kind of emotional attachment that a person can feel for an object. It is a kind of symbiotic love for an object that deserves affection not for what it does, but for what it *is*. Acknowledging the existence of *aichaku* in our built environment helps us to aspire to design artifacts that people will feel for, care for, and own for a lifetime.

THE ART OF MORE

In November of 2005 an exhibition of my digital art opened at the Fondation Cartier in Paris. Opening at the same time was a show of work by Australian artist Ron Mueck, a soft-spoken and intense man famous for his large-scale but incredibly life-like sculptures. The individual hairs, the shining eyes, the skin painted with veins—every detail is perfect.

So perfect that, as you approach one of Mueck's pieces, you ask yourself, "Is it real?" As your hand reaches out to confirm the warmth of the human form before you, your mind tells you that the sculpted giant cannot exist.

The best art makes your head spin with questions. Perhaps this is the fundamental distinction between pure art and pure design. While great art makes you wonder, great design makes things clear.

Sometimes, though, clarity alone is not the best design solution. At my opening in Paris, an old friend from Milan told me of a powerful socialite who was diagnosed with cancer. While she was still reeling from the shock of the news, her physician informed her of his ten-minute time limit for appointments. Even in her fragile state, she would have to leave, so that he could deliver similar messages to waiting patients. Here, the extremely efficient design of his communication system lacked any appreciation for the ambiguous dimensions of feelings—the stuff of art.

Afterwards, this brave woman came up with a solution that could bridge the gap between message and emotion. With five months left to live, she started a foundation to create intensely artful, beautifully designed centers near oncology units, where those first facing death can soak their minds and hearts. Art—a reason to live—is tempered with design—the clarity of message.

Achieving clarity isn't difficult. The Italian woman's oncologist had easily mastered it. The true challenge is achieving comfort.

Emotional intelligence is now considered an important facet of leaders today, and the expression of emotion is no longer considered a weakness but a desirable human trait to which everyone can immediately relate. Our society, systems, and artifacts require active engagement in care, attention, and feeling—the business value may not be immediately apparent. But the fulfillment from living a meaningful life is the ROE (Return on Emotion). A certain kind of more is always better than less—more care, more love, and more meaningful actions. I don't think I need to say anything *more* really.

LAW 8

Law 8

TRUST

In simplicity we trust.

Imagine an electronic device with only one unlabeled button on its surface. Pressing the button would complete your immediate task. Do you want to write a letter to Aunt Mabel? Go ahead and press the button. *Click*. A letter has been sent. You know with absolute certainty that it went out and expressed exactly what you needed. That's simplicity. And we are not far from that reality.

Every day the computer becomes increasingly smarter. It already knows your name, address, and credit card number. Knowing where Aunt Mabel lives and having watched you write a letter to her before, the computer can send a fair approximation of a kindly email to her from you. Just click a button and the deed could be done—finito. Whether the message is coherent and keeps you on dear Aunt Mabel's Christmas list is another story, but that is the price of not having to think. *In simplicity we trust*.

Hosting an email account on Yahoo! or MSN means that you can easily access your email from anywhere in the world. Another advantage is that the email service can customize itself based on your contact list and the kind of messages you send most often. For instance, a "send to Aunt Mabel" button can automatically appear just before her birthday. It is easy to forget, however, that the entire details of your e-social life are exposed to a company, or potentially a government, outside of your direct control.

The question is how comfortable you are about the computer knowing how you think, and then how tolerant you will be if (and when) the computer makes a mistake in guessing your desires. Most people would gladly give up some of the rote details of their life to have more free time, as expressed in the third Law. But is the risk of placing trust in the devices around you worth the simplicity gained? The issue of privacy in the digital age cannot be resolved in these next few pages, and thus we approach the issue of trust in a simpler manner.

RELAX, LEAN BACK

Learning how to swim as an adult was not easy. As an MIT undergraduate, I had managed to slip past the swimming requirement by showing that I could stand up in the pool. After leaving MIT, I tried all sorts of swimming programs to no avail. The return experience of learning how to swim at MIT was more successful. I admit that as a professor, taking swimming class with freshmen was kind of odd. I had just joined the MIT faculty and something about a swimming suit and goggles made me look more like an older student than a professor so I blended in quite well. I would be asked, "What major are you?" by the other students in the class. I kept my secret quiet.

My unorthodox swimming teacher did not teach us how to swim. He instead spent most of the term teaching us how to "lean back" and trust the water. I kept waiting to learn how to swim, but in the meantime became more comfortable just leaning back or bending over forward in the water. A formative moment occurred when he told us to go ahead and flap our arms and feet, and suddenly I was swimming! I realized I could always swim—I just didn't trust the water.

I was reminded of my swimming epiphany recently when I had the fortune of meeting the innovation director of Danish stereomaker Bang & Olufsen. As the Maserati of consumer electronics in style, attitude, and price, B&O struck me as an important data point in my search for understanding simplicity. Their legendary remote control (discussed in the first Law) embodies such qualities of simplicity as careful organization and attention to contrast. I was eager to engage in a discussion of simplicity that could help me understand the logic or, better yet, the *spirit* of the design philosophy that renders consumer electronics as high art. The answer, as it turned out, was simple.

B&O doesn't focus on the quality of sound, but on the quality of *leaning back* ... and just enjoying something. This was an unexpected lesson, but is consistent with the peripheral focus of the sixth Law. The goal of LEAN BACK is to achieve relaxation as the desired state, upon which audio and video can gradually invade, but not with intent to intrude. We can only truly relax when we trust that we're in the finest hands and are treated with the best intentions. A B&O system instills the same immersive trust that we grant to the water in the pool when we lean back and float.

Being able to lean back and relax often seems impossible in our competitive society. B&O's exquisite design inspires you to lower your guard. Their extraordinary attention to detail melts fear into safety—causing you to float away in its care.

That is, until your bliss is interrupted by your spouse and a no-no finger pointing to an outrageous charge on the credit card. The premium price of the B&O lean back experience is daunting, but consider that it is available at a lower price point in a park near you on a nice warm day where a bed of green grass has your name on it. Just lean back, for free.

TRUST THE MASTER

The power of the negative media around the food industry drives my mind to enact a Woody Allen-ish skit whenever faced with a restaurant menu. For instance, beef translates to "mad cow disease," chicken morphs into "avian flu," fish reconstitutes to "mercury poisoning," and vegetarian option becomes "genetically modified crops." I am unsure what to pick, and moreover who I can trust when my selection is made.

An alternative to such menu stress is available in better sushi restaurants where you can ask for the *omakase* (oh-maw-kaw-say) course. *Omakase* translates roughly to "I leave it up to you" where "you" refers to the sushi chef. The process is simple. The sushi chef looks at you, does a rough analysis of your general disposition, reflects upon the season and the day's weather, factors into consideration the variety of fish he has available in his arsenal, forms a rough idea for the optimal menu, starts the process of delivering the meal in measured increments, attentively observes your reaction, and tweaks the meal accordingly.

There is usually a fixed price for this special service by the sushi chef, but there is no shame in specifying your general cost parameters. The trick to the culinary satisfaction of *omakase* is not directly linked to cost but instead to the chef's confidence in his studied craft. This form of egotistical self belief is rooted in the Master's "manly pride," or *konjo* (kohn-joe)—which is likely more important than his own life, or at least so the lore of the Master goes.

The Western equivalent of *omakase* is the "chef's menu." From appetizer to main entrée to dessert, an exquisite choice of two or three options is offered each step of the way. Thus the chef's menu results in a great meal because the best dishes of the evening are put forward.

However, there are a few critical differences between the chef's menu and the *omakase* process. For example, the chef's menu is a lower risk approach because ultimately the blame for any mistake is on the diner for the choices they make for each course; the *omakase* approach is higher risk because the complete responsibility resides with the Master. Furthermore, in the chef's menu approach the cook is in the kitchen, far removed from the process of ordering, and unable to assess whether the meal offered will perfectly fit the needs of the diner. Instead in the *omakase* case, the diner sits only a few feet away from the sushi Master, and thus the Master's duel for winning the diner's tastebuds can have a life-or-death quality.

Vanity is a high risk sport that raises the stakes when all you can offer to a client is your word and your reputation as a

Master. Overconfidence is usually the enemy of greatness, and there's little room for personal ego when pleasing a customer is the true priority. But there's something to be said for the sushi Master's confidence. He knows with 100% accuracy that he will give a diner what she wants if she is willing to submit to his mastery and expertise.

Perhaps the *omakase* course is a form of subjection to culinary sadism—a gastronomic deviance that faces extinction in a progressively risk averse world. A sushi Master does not acknowledge risk; he has no fear. He has earned the trust of his customer, or else will fight literally with his bare hands to earn it when given the opportunity. Simplicity is achieved through the heroism of the trusted Master because in his sushi, we trust.

JUST UNDO IT.

It's the winter holiday season and you are buying a present for a friend. For each gift, you are issued a gift receipt that she can use to UNDO the purchase, exchanging it for a different one. Upon exchanging it, she can then be issued another receipt, with which she could exchange the gift again.

Knowing that a purchase is correctable later makes the shopping process simpler because you know that any decision made is not final. Indeed, today's customers don't expect to be held accountable for their purchases. Eager to build consumer trust in their brands, companies are willing to assume the extra risk inherent in a returnable purchase. The losses incurred by the cost of returned goods are outweighed by the gains in customer trust. This is the power of undo.

Computer tools give us the option to undo often, and now infinitely. Digital media is a forgiving media. Any visual mark, spoken utterance, or typed word entered into the digital domain can just as easily be removed. People have different opinions on the magic of undo. Some believe that the feature makes people more creative by allowing them to take more risks; others assert that undo makes people less creative because they don't think through ideas but rather create by happenstance. Which stance you take depends upon whether you are the sushi Master or just the average Joe.

From time to time I find myself romanticizing the old-fashioned typewriter and the messy little bottles of white correction fluid—the paper equivalent of undo. But a modern word processor is a comfort that I would be *an idiot* ... UNDO ... *remiss* to give up. A product that can correct our mistakes as they happen performs an important service and gains our trust. Undo is the welcome antidote to the average Joe's lack of optimism. In the end we can't all be the sushi Master.

The fourth Law of LEARN asserts the power of knowledge, which underlies the Master's ability to confidently execute any task without a crutch like undo. We trust that his skills are absolute and unerring—otherwise, why call him "Master" in the first place? Similarly, the self-assured design of a B&O stereo system allows you to LEAN BACK and relax in the care of the Master machine. Trusting a power greater than our own is a custom that is ingrained from birth when the adults that care for us provide the ultimate experience of simplicity. Every need and desire is met by a parent; and in return, beyond just offering our trust, we entrust our love.

On the contrary, undo is not about love, but simply a relationship of convenience. Power is equally balanced between experience and user such that neither side has the upper hand. There can be no relationship of depth because every interaction can be completely rewound to the beginning. Thus commitment is rendered meaningless when for every action, there is a corresponding un-action. In contrast to the trusting relationship with a Master, the power of undo results in a feeling of simplicity that is rooted in not having to care at all. Although there is something morally sad about this interpretation, undo is not the enemy. Embrace undo as a rational partner in maintaining the many complex relationships with the objects in your environment. But put the UNDO button away when dealing with real people if possible.

TRUST ME

As predicted in the third Law of TIME, Google's "I'm feeling lucky" button, which aims to take you to the single page you are looking for, will never be wrong and it will no longer need luck. Instead, Google will rely on its knowledge of your past habits to predict your current needs or desires. Searching for "soup"? You're probably searching for Campbell's soup because that's the soup most recently stocked in your smart pantry. Searching for "good book"? You probably will be looking for books similar to what you've purchased in the past. Amazon.com already has this suggestion engine, and although it isn't 100% accurate, increased computing power in the future will only aid machines trying to understand each of our particular quirks.

The more a system knows about you, the less you have to think. Conversely, the more you know about the system, the greater control you can exact. Thus the dilemma for the future use of any product or service is resolving the following point of balance for the user:

HOW MUCH DO YOU NEED TO KNOW ABOUT A SYSTEM?	⟷	**HOW MUCH DOES THE SYSTEM KNOW ABOUT YOU?**

On the left hand side, effort is required to learn and master the system; on the right hand side, trust must be offered to the system, and that trust must be consistently repaid. Privacy is sacrificed for extra convenience when following the Master's lead. Alternatively, undo allows us to become the Masters ourselves by gently learning to trust our own knowledge of a system. The placement of faith goes many ways.

On a final note on trust, years ago while in graduate school I had an officemate with a particularly cynical perspective. One day he warned me, "John, when someone says to you, 'Trust me,' replace every instance of that phrase with, 'F*ck you.'" Somebody asking you for their trust was, he believed, implicitly giving you the shaft. At the time I was the picture of naiveté, and afterwards I had difficulty UNDO-ing this naughty concept from my mind. For simplicity's sake, I've since learned to trust unquestionably in spite of my officemate's advice, but I am open to UNDO-ing that trust whenever deserved.

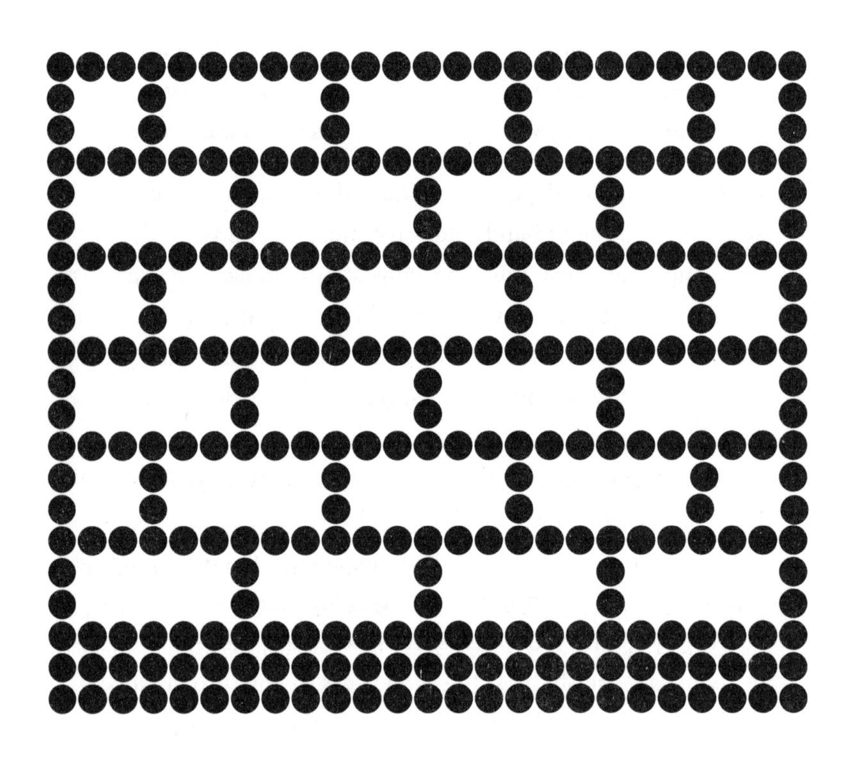

LAW 9

Law 9

FAILURE

Some things can never be made simple.

The truth embodied in the ninth Law is something I could have chosen to HIDE, but the eighth Law of TRUST commands me to speak. *Some things can never be made simple.* Knowing that simplicity can be elusive in certain cases is an opportunity to make more constructive use of your time in the future, instead of chasing after an apparently impossible goal. However there's no harm in initiating the search for simplicity even when success is deemed as too costly or otherwise out of reach.

There's always an ROF (Return On Failure) when you try to simplify—which is to learn from your mistakes. When faced with failure, a good artist, or any other member of the creative class, leverages the unfortunate event to radically shift perspective. One man's failed experiment in simplicity can be another man's success as a beautiful form of complexity. Simplicity and complexity shift with subtle changes in point of view.

Concentrate on the deep beauty of a flower. Notice the many thin, delicate strands that emanate from the center and the sublime gradations of hue that occur even in the simplest white blossom. Complexity can be beautiful. At the same time, the beautiful simplicity of planting a seed and just adding water lies at even the most complex flower's beginning. A relatively simple bit of computer code can produce surprisingly complex visual art. Conversely, Google's complex network of servers and algorithms produces a simple search experience. Deeming something as complex or simple requires a frame of reference.

There are certain things that I would never want to become simple—that includes my close relationships and my collection of art. Complexity and simplicity are two symbiotic qualities. As raised in the fifth Law of DIFFERENCES, each needs the other—its respective definition depends upon the other's existence. To realize a world of complete simplicity would mean that complexity would have to become completely eradicated. And with only simplicity remaining, how would you know what is truly simple? Thus failing to achieve simplicity is an important service to humanity.

Failure happens. If not 3.4 times out of a million, then at least one time today for you or me. I began my personal trek towards simplicity just at the turn of this century, and I am the first to admit that I do not have all the answers. Some of my thoughts will inevitably be deemed as wrong. But the impatience embodied by the third Law of TIME compels me to publish this book right now even with its unresolved flaws.

THE FLAWS OF SIMPLICITY 1: ACRONYM OVERLOAD

1 **REDUCE** The simplest way to achieve simplicity is through thoughtful reduction.

2 **ORGANIZE** Organization can make a system of many, appear fewer.

3 **TIME** Savings in time feel like simplicity.

4 **LEARN** Knowledge makes everything simpler.

In developing a methodology to support the first Law, I had a choice of SHE (SHRINK, HIDE, EMBODY) or HER (HIDE, EMBODY, REMOVE). Pronoun versus adjective is the first difference, and I thought of integrating the two parts of speech into the discussion. For instance, I played with being able to refer to HER and SHE interchangeably in the first Law's development. But it was the REMOVE in HER that made me remove HER in favor of SHE. Already I can see that I was correct to select only one, as this now sounds a bit like Abbott and Costello's famous "Who's on First?" comedy routine.

Later in the second Law of ORGANIZE I introduced SLIP (SORT, LABEL, INTEGRATE, PRIORITIZE), brought back SHE for the third Law, and then tried to discretely insert my BRAIN in the fourth Law of LEARN when I thought you weren't looking. Acronyms are a great way to simplify complex ideas, but the monotony of YAA (Yet Another Acronym) is too much to bear.

THE FLAWS OF SIMPLICITY 2: BAD GESTALTS

5 **DIFFERENCES** Simplicity and complexity need each other.

6 **CONTEXT** What lies in the periphery of simplicity is definitely not peripheral.

7 **EMOTION** More emotions are better than less.

8 **TRUST** In simplicity we trust.

As the Laws progress in the book, the themes become increasingly ambiguous. In the second Law I introduce the concept of gestalt—or the ability of the mind to "fill in the blank"—which justifies my approach to allow creative interpretation. However this open explanation can be confusing if taken logically.

The fifth Law of DIFFERENCES implies that there is a harmony between simple and complex that is achieved through human instinct. Everyone's instinct is different, and thus a single answer is not readily available to achieve the optimal balance between simplicity and complexity. For the same reason that there are a variety of musical styles like classical, rock, and hip-hop to satisfy differences in culture, curiosity, and fad, the rhythm of simplicity will be varied.

Next, in the sixth Law of CONTEXT I tell you to avoid the existing problem and to instead, look at the overall context of the situation. This approach may sound a bit irresponsible because it seems to imply that you should ignore the task at hand. Actually, the sixth Law doesn't suggest a path of direct neglect, but instead advocates concentrating on the invisible chasm that bridges the foreground task and its background context. However since this bridge I refer to is imperceptible, it doesn't seem fair for me to ask you to point your attention at what appears to be nothing. Also I imagine it doesn't help to say that "nothing is something" because it seems like I am making something out of absolutely nothing, which I am.

When emotions are a priority, and deep feelings come into play, I eschew the importance of complexity as delivered by pouring on more decoration, more glamour, and generally more flavor. Thus the seventh Law of EMOTION can be misinterpreted as saying that pure and simple experiences are sterile and devoid of feeling. It all depends on your personality and the mood that you wear at the exact moment of engagement. Sometimes you prefer clarity, and sometimes you prefer chaos. The seventh Law reserves your right to change your mind.

Finally in the eight Law of TRUST, I refer to the sushi Master as a persona worthy of absolute faith. Within almost the same breath, I espouse UNDO as the desirable power of not having to trust your own actions. Relieving yourself of pressure can feel fantastic, so why wouldn't a sushi Master want his own form of undo key sitting next to the sushi bar? Magnificent individuals in jobs that demand maximum performance of themselves tend to deny themselves the perceived weakness of the undo crutch, but it doesn't mean that they don't know how to relax. After all, that's what *sake*'s for.

THE FINAL FLAW: TOO MANY LAWS

9 **FAILURE** Some things can never be made simple.

When I initially set my goal on the Laws of Simplicity, I began with a target of sixteen—knowing that it was too many. After a few iterations of SLIP, I reduced the number to nine Laws which is in the attractive single digit category. Further integration of the Laws into a smaller set is feasible I suppose, but not necessary at this very moment because their evolution continues on the companion website *lawsofsimplicity.com.*

For the enjoyment of the simplicity purist that demands fewer guiding principles, I provide a single Law to remember as described in the following tenth Law: THE ONE.

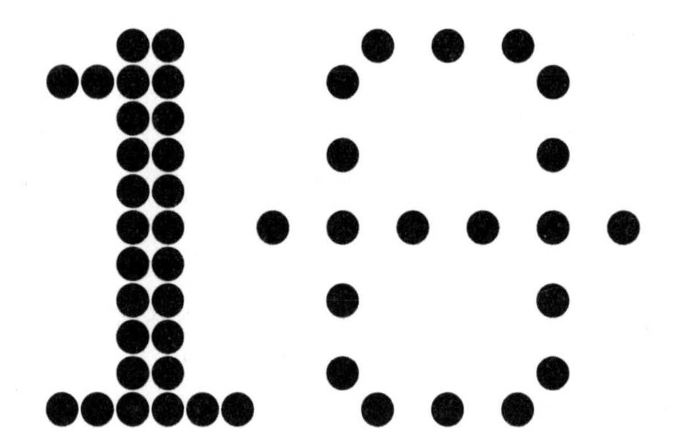

LAW 10

Law 10

THE ONE

Simplicity is about subtracting the obvious, and adding the meaningful.

The Japan National Rugby Team was once a mighty force that has fallen in recent years. Led by a new French coach, Jean-Pierre Elissalde, they appear to be on the rise. When Ellisalde first came aboard, he assessed the team's basic problem—the players were too predictable. As they moved up the field, the ball was passed between team members with a mechanical accuracy that was easy for their opponents to predict, and thus consistently topple. Elissalde urged his players "to become like the bubbles in a glass of champagne," floating upward in unexpected and elegantly fluid ways. The Japanese team had to learn how to operate based upon intuition versus intellect.

Simplicity is hopelessly subtle, and many of its defining characteristics are implicit (noting that it hides in SIMPLICITY). Drinking deeply from Ellisalde's champagne approach led me to a single, simplified expression: *Simplicity is about subtracting the obvious, and adding the meaningful.*

Ten laws (10: *one, zero*), remove none (0: *zero*), and you're left with one (10: *one*). When in doubt, turn to the tenth Law: THE ONE. It's simpler that way.

After SLIP-ping my observations into the ten Laws of Simplicity, I found that several ideas didn't fit neatly into any single Law. They did, however, cluster around three specific technologies with particular relevance to the subject of simplicity. Originally, I thought of REDUCE-ing the book by removing these three sections. But in discussions with a variety of business leaders I felt that they weren't completely obvious so by THE ONE Law I have kept them here.

Key 1
AWAY

More appears like less by simply moving it far, far away.

I cannot forget the moment, on a cold New England night in 1984 in the comfort of a friend's dorm room, when I watched him type some magical incantation into the computer terminal that allowed him to jump from a mainframe computer at MIT to another mainframe at Columbia University. "No way!" I said. His steely reply was a monotonic Keanu Reeves-ish, "Yes way."

Because the university's big central computers were more powerful than the then new personal computers, many of the tech-savvy students opted for lower cost data terminals—a text display with no computational power of its own but the ability to connect to more powerful machines. There was a kind of macho-ness to having less on your actual physical desktop, but being able to do more remotely.

Desktop computers today have as much processing power as that central MIT mainframe we jacked into decades ago. Yet with less than one percent of the average desktop computer's processing capabilities, your basic word processing and spreadsheet applications can run comfortably. Despite this fact, with so much memory and horsepower available, today's applications have become bloated. What could once be installed from a single floppy disk grew to fill an entire CD, then a set of CD's, then a DVD, and now multiple DVD's.

When these supersized tanks of data are poured into the computer, the equivalent of an accidental oil spill is likely to occur in the ocean of virtual information. The result is a computer that is no longer spry as the day it was unpackaged, or in the worst case it can't even start up. Maintaining an up-to-date computer can feel like a full-time job for its owner.

A revolution is occurring that looks a bit like a devolution—the simple model of the data terminal is regaining popularity not for its macho-coolness aspects, but for its appeal to common sense. Rather than deal with a stack of CD's or network downloads to keep the computer on your desk going, why not simply access software on a remote computer?

Think of the power of Google which runs from a simple, lightweight text input box in your web browser to access Google's vast network of computers and databases. You are spared having to house your own massive racks of computing equipment required to process a Google query. *More appears like less by simply moving it far, far away.* Thus an experience is made simpler by keeping the result local, and moving the actual work to a far AWAY location.

This model of computer applications running remotely is gaining popularity and is called "software as a service." Google is free (for now), but one could imagine it as a future service whereby we pay per query or on a monthly basis because of the value received. Don't forget the convenience of not having to maintain or manage the computational horsepower to run the software locally. Already business-focused software systems for running spreadsheets, managing projects, and maintaining customer relationships, like the popular Salesforce.com, are available as services on the Web. Not only do these systems feel simpler by being hosted far away, but they also importantly acknowledge the fact that we live in a mobile world where we're often away from the office or home.

Fundamental to the effectiveness of AWAY is how to maintain reliable communication with an outsourced task. A web-enabled phone is only good when it can reliably access the network. Conversely, a remotely hosted service needs to be resistant to the latest virus or hacker attack. It is comforting to think that even in the 21st century, the question of how to maintain a long distance relationship continues to flourish.

Key 2
OPEN

Openness simplifies complexity.

Being truly OPEN in our open society can be risky business. People routinely risk emotional pain when they expose themselves with the simple words, "I love you." When the response is positive, the angels sing and fairies dance in the air; when the response is negative, the angels and fairies have left town to never come back. In the parlance of the business world, professing your love for someone is a high risk, high reward opportunity. As a person happily engaged in a relationship that has lasted for more than fifteen years now, I'm glad to have taken the risk.

Companies don't tend to profess love in the same way, but there is increasing pressure on businesses to design products to be more OPEN. Opening a proprietary system, much like professing one's love, is a high risk activity that a company posting quarterly-earning figures often cannot afford. Who might misuse the information? What if our competitors leverage our company secrets? Why would a consumer buy what they could now easily make themselves? Giving away what is perceived to be the core protectable value—i.e. know-how, or "intellectual property"—does not make sense when tremendous efforts and investments have gone into realizing a successful product.

In the technology world, the "open source" model—in which source code, the equivalent of a software's blueprints, is made publicly available—is championed as a way to generate software that is not only free, but more robust than most software available on the market. The best-known example is Linux, an operating system that competes with Microsoft Windows. While Linux is free and open source, Windows is for-pay and closed source.

I once heard a Linux expert on the radio explaining that when Windows is broken you cannot fix it yourself because the source is closed, whereas with Linux you can. This is fairly misleading, actually, because as computer programs go, Linux is extremely complex. Even with access to the code, your average computer user wouldn't be able to fix a bug. That requires an expert. However, there are thousands of Linux experts on the Net at any time that can respond to common problems such as security flaws. These experts are more likely to jump into action before you'd even get to a real Microsoft employee on the phone. *Openness simplifies complexity.* With an open system, the power of the many can outweigh the power of the few.

A second model of open source that is more palatable to businesses not wanting to give away their source code is to offer an Application Programming Interface, or "API." Amazon.com was an early pioneer of this approach—offering open access to its running components, instead of the actual source code, through the Amazon.com API. This API enables any person on the Web to design and build her own book store. Another example is the Google Maps API that lets other programmers build new apps like a route planner for runners or a real estate map.

An API is thus a selective approach to open systems where the functionality, instead of the actual blueprints as in open source, is offered to the general community to the extent that excess processing capacity can be made available. Note that this functionality is usually offered to the community free of charge.

According to the eighth Law, a deep form of simplicity is rooted in TRUST. Any book on salesmanship will tell you that trust forms the basis of a strong business relationship. Open systems place unique demands on the economics of trust. If the adage, "it is better to give than receive," rings true to you, then the long run gains associated with an open system will also be obvious to you. If conventional capitalism is your compass, and to hear "trust me" translates to "f*ck you," then you will likely choose the closed approach. However, there are signs that a "for free" open approach can lead to a "for a fee" approach. For example, the popular "Ruby on Rails" Web framework by 37signals is completely free, but related for-pay services are sold simultaneously. The case on OPEN is open, indeed.

Key 3

POWER

Use less, gain more.

Every rechargeable device I own is like a new pet that must be fed. The magic of cordless systems such as mobile phones, laptops, and so forth is freeing, yet there is a toll exacted with each new device acquired. I know that if I do not feed each device with energy regularly, batteries begin to discharge and their efficacy will eventually fade.

I own an iPod but I never really listen to music anymore as usually I like to listen to the sounds around me. It sits on my desk and I may turn it on once every few weeks only to realize its battery is discharged. With the odd, ritual feeling of managing a critically ill patient, I rush to connect the little fellow up to the power dongle, and feel relieved when a pulse is visibly returned. But I know in the back of my mind that one day it will not revive from its deep sleep due to the finite nature of rechargeable battery technology. We wear out as humans, so it's only fair and natural that batteries should wear out too.

My colleague Prof. Joseph Paradiso is developing new solutions to the problem of POWER. He and his team at MIT have invented a self-powered, wireless switch that harvests the energy generated during the push of a button to electrically send a radio-frequency signal. Said another way, the key fob that activates your car alarm system will not need a battery and instead will use just the power recovered from your pushing of the button. It's just a tiny handheld switch, but it's arguably one of the most popular inventions at the Media Lab. A similar workaround for battery life is seen in extremely low-power electronic circuitry that enables certain devices to last on a single battery for decades. Electronic devices can never be truly simple unless they are freed from their dependence on power. A seemingly unpowered electronic device may seem like an oxymoron, but it is critical to achieve.

The US is at a turning point in its development. The mercurial cost of fuel and its inevitable link to geopolitics make any discussion of power complex. We need it, and with the continually growing world population we'll always want and need more. A rechargeable battery, or any battery technology for that matter, has the guise of freedom—it seems to free you from dependence on an external power. But all power comes from somewhere and uses energy on its way to the consumer—batteries must be manufactured, ditto with solar panels, oil must be transported across great distances. The only foreseeable solution is for humanity to collectively use less energy, and to use it more wisely. *Use less, gain more.* A personal sacrifice can directly translate to a philanthropic act for the world that although not tax deductible, makes simple sense.

I practice my own kind of "sustainable computing." In recent times I have begun to play a businessman's equivalent to the daring game of "chicken" where I see how much life I can get out of my laptop on a trip without bringing the power cord. In the field of design there is the belief that with more constraints, better solutions are revealed. With only 14 minutes of charge left on my laptop right now, I find that indeed much more can get done than when the power is fully connected and freely available. Urgency and the creative spirit go hand in hand, and innovation as a positive return is a desirable benefit. The number of people who will see the benefit of this approach will determine the terminal point on the progress bar of our glorious planet Earth. Increased social practices that result in the use of less power—as well as supporting technology innovations for power harvesting and conservation—stand to realize a world where the most powerful examples of simplicity are those that will ironically appear powerless.

The three Keys of AWAY, OPEN, and POWER are important technology markers for the future of simplicity. Openly discussing and debating the three Keys, and more Keys to come, continues on the *lawsofsimplicity.com*.

Technology and life only become complex if you let it be so.

While drawing with pen and paper in art school, and reaching for the nonexistent UNDO key to correct a mistake, I began to feel that technology was shaping me more than the other way around. Around the same time, a friend told me about the thinker, Ivan Illich, and his writings on how the emergence of professions has disabled the average person. Lawyers solve problems between people that in the past we resolved ourselves; doctors cure people, whereas in the past we knew which plants in the forest had medicinal properties. The lesson I've taken from Illich's work is that while technology is an exhilarating enabler, it can be an exasperating *disabler* as well.

For instance, I recall waiting for several days to get a refill for my label printer when it occurred to me that I could just write on the file folder with a pen. Or, whenever there's a question about an unknown word my first instinct is to go to dictionary.com. But by the time I've awakened my computer to type it in, someone in my house has found it by flipping through an actual dictionary. I have stood nervously in front of an audience of hundreds of people held up while my computer unsuccessfully talks with the data projector; I then remember that I do a better job presenting ideas without PowerPoint. The disabling effect of technology can be humorous in retrospect. But sometimes I wonder if being a Blackberry-toting cyborg is all that it's cracked up to be.

Every day some of the smartest young people in the world come to see me in my office at MIT. Although officially I am their teacher, I find that I am often their student. For instance, I remember a student named Marc who volunteered in shelters for poor people at the end of their lives. Even though he came from a well-heeled family and could easily turn his back on the impoverished, Marc said he always felt compelled to help others in need. He told me how while working at the shelter, he noticed that each patient had a single shelf by their bed that held the total sum of their worldly belongings. Seeing this situation made him silently ask, "What are the few precious things that you can afford to keep at the end of your life when you already have so little?" A ring, a photograph, or another small memento was what he consistently found. Marc poignantly surmised that memories are all that matter in the end.

When your entire life is reduced to a single shelf of curios, what memories might you enshrine? Life may be complex, but in the end, life is simple if you listen to Marc.

The ten Laws and three Keys are not the end of my thoughts about simplicity. Encouraged by those with whom I have shared these thoughts so far, I plan to continue this mission. MIT Press has other titles to come in this series on simplicity. The next installment—*The Value of Simplicity* by the stunningly insightful Jessie Scanlon—will take a modern business focus. If you would like to join the emerging discussion, please visit *lawsofsimplicity.com*. I promise to keep it simple.

TEN LAWS

1. **REDUCE** The simplest way to achieve simplicity is through thoughtful reduction.
2. **ORGANIZE** Organization makes a system of many appear fewer.
3. **TIME** Savings in time feel like simplicity.
4. **LEARN** Knowledge makes everything simpler.
5. **DIFFERENCES** Simplicity and complexity need each other.
6. **CONTEXT** What lies in the periphery of simplicity is definitely not peripheral.
7. **EMOTION** More emotions are better than less.
8. **TRUST** In simplicity we trust.
9. **FAILURE** Some things can never be made simple.
10. **THE ONE** Simplicity is about subtracting the obvious, and adding the meaningful.

THREE KEYS

1. **AWAY** More appears like less by simply moving it far, far away.
2. **OPEN** Openness simplifies complexity.
3. **POWER** Use less, gain more.

There are a few books that inspired each of the sections that I owe the debt of inspiration to mention here. I omit the practice of listing a bibliographic entry for each item, because the Web has made it simple to find a book so why make it look complex?

SIMPLICITY = SANITY

The Tipping Point, by Malcolm Gladwell (2002)
The need for simplicity has reached the tipping point.

REDUCE

The Paradox of Choice, by Barry Schwartz (2005)
Provides a grounding in why few can be better than many.

ORGANIZE

Notes on the Synthesis of Form, by Christopher Alexander (1964)
Ideas about organization as originated in architecture.

TIME

Toyota Production System, by Ohno Taiichi (1988)
Dry treatise on optimizing production from the Toyota Master.

LEARN

Motivation and Personality, by Abraham Maslow (1970)
What really motivates people?

DIFFERENCES

The Innovator's Solution, by Clay Christensen (2003)
Simple explanation of changeover effects led by technology.

CONTEXT

Six Memos for the Next Millennium, by Italo Calvino (1993)
Brilliantly beautiful thoughts on simply everything.

EMOTION

Emotional Design, by Donald Norman (2003)
Usability guru makes a case for the useless.

TRUST

The Long Tail, by Chris Anderson (2006)
Adding up all the little things really matters.

AWAY

Technics and Civilization, Lewis Mumford (1963)
Prescient work by a man in touch with his time.

OPEN

The Wisdom of Crowds, by James Surowiecki (2004)
Supports the group outweighing the individual.

POWER

Cradle to Cradle, by W. McDonough and M. Braungart (2002)
We're running out of power and something has to be done.

LIFE

Disabling Professions, by Ivan Illich (1978)
Reminds you that you're becoming increasingly useless.

MAEDA@MEDIA (2001) and *Creative Code* (2004) document my own creative genesis.

INDEX

I read a scathing review on Amazon.com for a book that did not include an index and also did not include complete references for each factoid presented. For LOS, I made a conscious choice to not make a book that is a compendium of facts because I don't feel comfortable with managing that kind of complexity. An index, on the other hand, I can handle.

FEBRUARY 2, 2005

I used to see an older fellow at the MIT pool almost every day. He was, he told me, a retired professor of linguistics.

Today I saw him in the locker room after a long hiatus, and we had a brief conversation about "insecurity," a topic that I'd been thinking about.

"The thing with insecurity, is that if you are too insecure, then you don't grow—because you're paralyzed by the fear of failure," I said to him, out of the blue. "On the other hand, if you have no insecurity, then you don't grow either—because your head is so big that you can't recognize your failures."

"Balance in all," the professor emeritus replied.

Then I posited, "If you are in the middle, however, you have to shift towards the edges and oscillate a bit in order to know if you are centered."

"You can get lost in the middle sometimes," he said.

We both fell quiet and I finished packing my things. Then, I was tying my shoes when I blurted, "Mentors."

The professor emeritus said in a firm voice, "You need mentors to give you courage."

I then sorrowfully parried, "But all your mentors tend to go away as you age."

The professor emeritus paused, and then responded, "Yes, because you don't need them anymore."

I shook his hand and said, "Thank you for the lesson." The Master professor smiled as he put his socks and shoes on, and I left the locker room thinking, "Exercise is truly good for the heart."

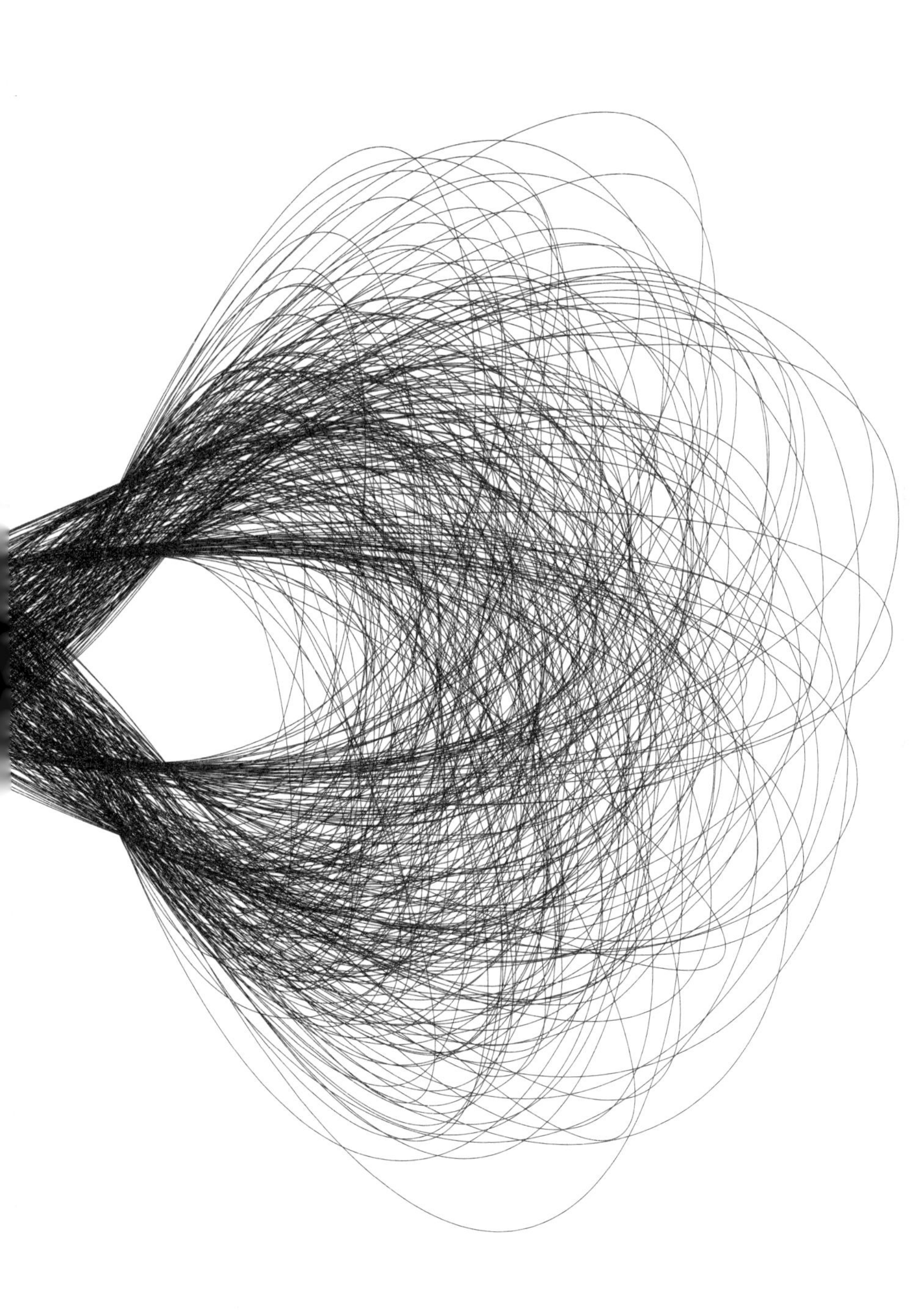